隧道与地下工程领域
融合创新精品教材

Underground Engineering
Ventilation Disaster Prevention
&
Environmental Control

地下工程
通风防灾与环境控制

郭 春 杨 璐 刘一杰 / 编著

人民交通出版社股份有限公司
北 京

内 容 提 要

针对公路隧道、铁路隧道、地铁等地下工程交通设施，本书对施工、运营阶段地下工程通风防灾的知识内容进行了归纳梳理，重点介绍了地下工程通风的相关卫生标准、通风方法、计算流程和设备布置，说明了各类地下空间的防灾救援体系，包括救援设施、救援策略、救援计划。本书适用于一般高校地下工程、城市地下空间工程方向的土木工程专业本科生和研究生。也可作为地下工程设计、施工和运行管理技术人员的参考。

图书在版编目(CIP)数据

地下工程通风防灾与环境控制 = Underground Engineering Ventilation, Disaster Prevention & Environmental Control：英文 / 郭春编著. — 北京：人民交通出版社股份有限公司，2020.7

ISBN 978-7-114-15819-3

Ⅰ.①地… Ⅱ.①郭… Ⅲ.①地下建筑物—通风—研究—英文②地下建筑物—防灾—研究—英文③地下建筑物—环境控制—研究—英文 Ⅳ.①TU96

中国版本图书馆 CIP 数据核字(2019)第 184882 号

Dixia Gongcheng Tongfeng Fangzai yu Huanjing Kongzhi

书　　名：地下工程通风防灾与环境控制
　　　　　Underground Engineering Ventilation, Disaster Prevention & Environmental Control
著 作 者：郭　春
责任编辑：王　霞　张博嘉
责任校对：孙国靖　宋佳时
责任印制：刘高彤
出版发行：人民交通出版社股份有限公司
地　　址：(100011)北京市朝阳区安定门外外馆斜街 3 号
网　　址：http://www.ccpcl.com.cn
销售电话：(010)59757973
总 经 销：人民交通出版社股份有限公司发行部
经　　销：各地新华书店
印　　刷：北京印匠彩色印刷有限公司
开　　本：787×1092　1/16
印　　张：14
字　　数：425 千
版　　次：2020 年 7 月　第 1 版
印　　次：2020 年 7 月　第 1 次印刷
书　　号：ISBN 978-7-114-15819-3
定　　价：68.00 元

(有印刷、装订质量问题的图书由本公司负责调换)

Preface

Some objective factors are accelerating the development and utilization of underground space: ground traffic congestion, lack of space, environmental pollution and energy consumption. Common underground transportation facilities include highway tunnel, railway tunnel, underground parking lot and metro. Besides, underground space also includes underground shopping malls, civil air shelters, parking lots and other facilities.

The construction ventilation system is an indispensable during the construction period of the underground storage, basement and other underground projects. construction ventilation is necessary to supply fresh air, remove dust and various toxic and harmful gases. It helps to create a pleasant working environment and ensure the health and safety of construction personnel. Natural ventilation is clean and renewable, but generally not sufficient for long tunnels. Mechanical ventilation consumes electricity and supplies abundant fresh air for the underground space. The ventilation arrangements are complicated when involving with multiple working faces or service galleries. But the basis of construction ventilation still lies in the calculation of air volume and ventilation resistance.

Underground space is a relatively closed space, so operation ventilation is an important technical link to ensure the health and safety of personnel in the underground space. The book elaborates on the operation ventilation of highway tunnels, railway tunnels and metro. Air conditioning ventilation is one important part of metro ventilation which differs from tunnel ventilation. Air conditioning system brings fresh air and balances the thermal load. Besides, the characters of the operation ventilation for underground shopping malls, parking lots, mines and civil air shelters are briefly introduced.

Fire accidents in underground space account for almost one-third of the total number of accidents. Limited to the long and narrow structure, the underground structure is easy to be at a disadvantage when facing a fire threat. The capability of fume exhausting and heat dissipation for underground space is limited. The air temperature and the concentration of toxic fume may increase quickly in finite space, which makes the evacuation and disaster relief difficult. The ventilation under fire disasters varies from common ventilation arrangement. The ventilation system should consider the unfavorable effect of fire fume and blocked traffic. The ventilation needs to provide fresh air for rescue channels. Different rescue strategies are introduced according to the characters of the underground space.

From the perspective of book content, the book contains three parts. The first part, chapter 1, focuses on the construction ventilation of underground engineering, it introduces relevant hygienic standards, ventilation methods as well as calculation processes and equipment arrangement of the underground ventilation. The second part, chapter 2-5, contains the operation ventilation of highway tunnels, railway tunnels, metro and other underground spaces, it specifies the calculation of air demand and ventilation method selection. The third part, chapter 6-9, describes the fire prevention and rescue of underground space which includes highway tunnels, railway tunnels, metro and other underground spaces. The rescue system contains the rescue facilities, rescue strategies, rescue plans. Fire control technologies include fire alarm system, the ventilation arrangement when the underground space is on fire. Readers should pay attention to the similarity and difference of the ventilation and fire prevention for different underground spaces.

In summary, this book provides a comprehensive overview of the various types of engineering in the field of underground engineering, such as highway tunnel, railway tunnel and metro. The essential ventilation and disaster prevention contents in the whole life cycle of design, construction and operation are combed and summarized. The ventilation and disaster prevention system of underground engineering is improved by combining the existing knowledge contents with the author's previous scientific research achievements. This book is suitable for undergraduates and postgraduates majoring in civil engineering with the direction in underground engineering and urban underground space engineering in general colleges and universities. It can also be used as a reference for technical personnel of underground engineering design, construction and operation management.

In the process of compiling this book, the editors have absorbed the advantages of many previous textbooks and consulted the scientific and technological literatures published in recent years at home and abroad. I would like to express my gratitude to the authors of these literatures. Although we have made great efforts, but due to the limited level of knowledge, omissions are unavoidable. Please point out the mistakes in the book by all means.

Author
January 2019

Contents

Chapter 1 Construction Ventilation of Underground Engineering Projects 1
  1.1 Hygienic Standard in Underground Engineering Construction Environment 1
  1.2 Natural Ventilation 7
  1.3 Basic Mechanical Ventilation 13
  1.4 Ventilation Methods in Common Tunnels with Access Adits 16
  1.5 Calculation of Construction Ventilation 35
  1.6 Construction Ventilation Equipment and Selection 42
  Exercise 51

Chapter 2 Operation Ventilation of Highway Tunnels 53
  2.1 Hygienic Standard for Operation of Highway Tunnels 53
  2.2 Calculation of Air Demand 55
  2.3 Ventilation Methods and Selection 59
  Exercise 71

Chapter 3 Operation Ventilation of Railway Tunnels 73
  3.1 Hygienic Standard for Operation of Railway Tunnels 73
  3.2 Air Demand Calculation 74
  3.3 Ventilation Methods and Selection for Tunnel Operation 76
  Exercise 82

Chapter 4 Metro Ventilation and Air Conditioning 84
  4.1 Overview of Metro Ventilation and Air-conditioning System 84
  4.2 Composition of Metro Ventilation and Air-conditioning System 86
  4.3 Internal Ventilation System of Metro Station 93
  4.4 Operation Status of Metro Ventilation and Air Conditioning System 99
  4.5 Calculation of Metro Ventilation and Air Conditioning System Load 101
  Exercise 105

Chapter 5 Operation Ventilation for Other Underground Space 107
  5.1 Operation Ventilation of Underground Shopping Malls 107

5.2　Operation Ventilation of Underground Parking Lot …… 111
5.3　Mine Operation Ventilation …… 114
5.4　Operation Ventilation of Civil Air Defense Works …… 120
Exercise …… 123

Chapter 6　Disaster Prevention and Rescue of Highway Tunnels …… 124
6.1　Fire Characteristics of Highway Tunnels …… 124
6.2　Standard for Calculation of Fire Ventilation Pressure and Fume Control in Highway Tunnels …… 128
6.3　Disaster Prevention and Rescue System for Single Tunnel …… 139
6.4　Disaster Prevention and Rescue System of Tunnel Group …… 149
6.5　Fire Prevention Technology of Highway Tunnel …… 152
Exercise …… 160

Chapter 7　Railway Tunnel Disaster Prevention and Rescue …… 161
7.1　Development of Disaster Prevention and Rescue in Railway Tunnels …… 161
7.2　Rescue and Evacuation Facilities …… 162
7.3　Ventilation Design for Tunnel Disaster Prevention …… 165
7.4　Ventilation Equipment Arrangement …… 168
7.5　Design of Wind Speed and Volume for Disaster Prevention and Rescue …… 170
7.6　Cooperation of Ventilation with Other Specialties …… 172
Exercise …… 173

Chapter 8　Metro Disaster Prevention and Rescue …… 174
8.1　Overview of Metro Fires …… 174
8.2　Automatic Fire Alarm for Metro …… 178
8.3　Emergency Evacuation Passage Technology for Metro …… 186
Exercise …… 189

Chapter 9　Disaster Prevention and Rescue of Other Underground Space …… 191
9.1　Disaster Prevention and Rescue of Underground Shopping Mall …… 191
9.2　Disaster Prevention and Rescue of Underground Parking Lot …… 196
9.3　Mine Disaster Prevention and Rescue …… 200
9.4　Disaster Prevention and Rescue of Civil Air Defense Engineering …… 205
Exercise …… 209

References …… 210

Chapter 1　Construction Ventilation of Underground Engineering Projects

[Important and difficult contents of this chapter]
(1) Hygienic standard of underground engineering construction environment.
(2) Natural air flow and common natural ventilation.
(3) Basic mechanical ventilation.
(4) Ventilation of common tunnels and access adit.
(5) Calculation method of construction ventilation, air leakage of line network and ventilation resistance.

1.1　Hygienic Standard in Underground Engineering Construction Environment

Tunnel construction generates harmful substances, namely dust and harmful gases, which pollutes the tunnel air and damaging the health of the workers inside the tunnel. These harmful substances can be divided into three categories: gas, dust and noise. Common harmful gases mainly include carbon monoxide, carbon dioxide, nitrogen monoxide, nitrogen dioxide, sulfur dioxide, hydrogen sulfide and gas.

1.1.1　Industry Health Standards

In order to protect the health of underground construction workers and ensure the safety of production, some hygienic standards of underground construction are proposed.

1. Hygienic standards stipulated in the current "Code for Railway Tunnel Construction"

At present, the railway tunnel construction is carried out according to *code for Railway Tunnd Construction* (*TB 1003—2016*) issued by the Ministry of Railways (Now the China State Railway Group co. ,Ltd), which clearly stipulates the oxygen content, dust concentration, harmful gas concentration, temperature and noise of the air in the tunnel, and requires that the working environment during the tunnel construction should meet the following standards:

(1) The oxygen content in the air shall not be less than 20% by volume.
(2) The allowable concentration of dust, which contains more than 10% of free silicon

dioxide per cubic meter of air, shall not be greater than 2 mg.

(3) Maximum permissible concentration of harmful gases: the maximum permissible concentration of carbon monoxide is 30 mg/m^3. Under special circumstances, when the construction personnel must enter the working face, the concentration is 100 mg/m^3, but the working time shall not be more than 30 minutes; Carbon dioxide shall not exceed 0.5% by volume; Nitrogen oxides (converted to NO_2) are less than 5 mg/m^3.

(4) The temperature in the tunnel shall not be higher than 28℃.

(5) Noise in the tunnel shall not be greater than 90 dB.

2. Hygienic standard stipulated in the current "Technical Specifications for Highway Tunnel Construction"

At present, the highway tunnel construction is carried out in *Technical Specification for Highway Tunnel Construction* (*JTG F60—2009*) issued by the Ministry of Transport in September 2009, which clearly stipulates the oxygen content, dust concentration, harmful gas concentration, temperature and noise of the air in the tunnel, and requires that the working environment during the tunnel construction should meet the following standards:

(1) The oxygen content in the tunnel air shall not be less than 20% by volume.

(2) The temperature in the tunnel should not be higher than 28℃.

(3) Noise shall not be greater than 90 dB.

(4) Dust concentration. Dust containing more than 10% free silicon dioxide per cubic meter of air shall not be greater than 2 mg.

(5) Harmful gas concentration: carbon monoxide is generally not more than 30 mg/m^3, and 100 mg/m^3 when the construction personnel must enter the working face under special circumstances, but the working time shall not exceed 30 minutes; Carbon dioxide shall not exceed 0.5% by volume; Nitrogen oxides (converted to NO_2) are less than 5 mg/m^3.

(6) In gas tunnel blasting, the gas concentration in air flow must be less than 1.0% within 20 m of the blasting site; the gas concentration in the air flow of the total return duct is less than 0.75%; all personnel must be evacuated to a safe place when the gas concentration in the excavation face is greater than 1.5%.

3. Hygienic standards stipulated in the current "Safety Regulations for Coal Mines"

The current "Safety Regulations for Coal Mines" were deliberated and adopted at the 13th Director-General's Office Meeting of the State Administration of Work Safety on 22nd December 2015 and came into force on 1st October 2016. The hygienic standards stipulated in this regulation are:

(1) In the intake air flow of the mining face, the oxygen concentration shall not be less than 20% and the carbon dioxide concentration shall not be more than 0.5%.

(2) The concentration of harmful gases shall not exceed the provisions of Table 1-1.

(3) The allowable concentrations of methane, carbon dioxide and hydrogen shall be in

Chapter 1 Construction Ventilation of Underground Engineering Projects

accordance with the relevant provisions of these regulations.

Maximum allowable concentration of harmful gas　　　Table 1-1

Harmful gas	Maximum allowable concentration (%)
Carbon monoxide(CO)	0.0024
Nitric oxide(Converted to nitrogen dioxide NO_2)	0.00025
Sulfur dioxide (SO_2)	0.0005
Hydrogen sulfide(H_2S)	0.00066
Ammonia(NH_3)	0.004

4. Hygienic standards stipulated in the current "*Safety Regulations for Metal and Non-Metal Mines*"

The current *Safety Regulations for Metal and Non-metal Mines* (GB 16423-2006) were promulgated by the State Administration of Work Safety in June 2006. The hygienic standards for underground mines stipulated in the Regulations are:

(1) The air composition (calculated by volume) in the intake air flow of the undergroundmining face shall be not less than 20% of oxygen and not more than 0.5% of carbon dioxide.

(2) The dust content of the air source in the air entry shaft and working face shall not exceed 0.5 mg/m³.

(3) The exposure limits of hazardous substances in the air entering and leaving the work place shall not exceed the provisions of *GBZ 2.1—2007 Occupational exposure limits for hazardous agents in the workplace. Part 1:chemical hazardous agents*.

(4) In mines containing radioactive elements such as uranium and thorium, the concentration of radon and its daughters in underground air shall conform to the provisions of *GB 4792*.

5. Hygienic standards stipulated in the current "*Safety Regulations for Metallurgical Underground Mines*"

The Safety Regulations for Metallurgical Underground Mines issued in April 1990 stipulate the following hygiene standards for underground mines:

(1) The air composition and oxygen content in the intake air flow of the underground mining face shall not be lower than 20% by volume, and the carbon dioxide content shall not be higher than 0.5% by volume.

(2) The dust content in the air of all operation sites under the shaft shall not exceed 2 mg/m³, and the dust content in the air source of the air entry shaft, roadway and mining face shall not exceed 0.5 mg/m³.

(3) The concentration of toxic and harmful gases shall not exceed the provisions of Table 1-2 at the downhole operation site (mine without diesel equipment).

(4) The concentration of toxic and harmful gases in the mine and downhole operation site

using diesel engine equipment shall conform to the following provisions: carbon monoxide less than 60 mg/m³, nitric oxide less than 10 mg/m³, formaldehyde less than 6 mg/m³, acrolein less than 0.6 mg/m³.

(5) The air temperature of the working face shall not exceed 27℃; The air temperature in hot water type mines and high sulfur mines shall not exceed 27.5 ℃.

(6) The dust concentration in the air of the workplace shall conform to the relevant provisions of *Design Hygiene Standard for Industrial Enterprises* (*GBZ1—2010*).

Maximum allowable concentration of hazardous gases in metallurgical mines   Table 1-2

Harmful gas	Maximum allowable concentration (mg/m³)	Gas	Maximum allowable concentration (mg/m³)
Carbon monoxide (CO)	30	Hydrogen sulfide (H_2S)	15
Nitric oxide (NO, Converted to nitrogen dioxide NO_2)	5	Ammonia (NH_3)	10

1.1.2 National Health Standards

The national health standards mainly include *Design Hygiene Standard for Industrial Enterprises* (*GBZ 1—2015*), *Occupational Exposure Limits for Harmful Factors in Workplaces Part* I: *Chemical Harmful Factors* (*GBZ 2.1—2007*) and *Occupational Exposure Limits for Harmful Factors in Workplaces Part* II: *Physical Factors* (*GBZ 2.2-2007*).

Allowable concentrations of hazardous gases in the workplace are included in the *Occupational Exposure Limits for Hazardous Factors in the Workplace Part I: Chemical Hazards* (*GBZ 2.1—2007*), which specifies the allowable concentrations of 339 chemicals and 47 dusts. The allowable concentrations of several chemicals (harmful gases) related to the tunnel construction environment are shown in Table 1-3, and the allowable concentrations of dust are shown in Table 1-4.

Allowable concentrations of chemicals in the air of tunnel workplaces   Table 1-3

No.	Name	Chemical abstracts (CAS No.)	OELs (mg/m³)		
			MAC	PC-TWA	PC-STEL
1	Carbon monoxide	630-08-0			
	Non-plateau		—	20	30
	Plateau				
	Altitude 2000-3000m		20	—	—
	Altitude >3000m		15	—	—
2	Nitric oxide	10102-43-9	—	15	—
3	Nitrogen dioxide	10102-44-0	—	5	10
4	Sulfur dioxide	7446-09-5	—	5	10
5	Hydrogen sulfide	7783-06-4	10	—	—
6	Carbon dioxide	124-38-9	—	9000	18000

Chapter 1 Construction Ventilation of Underground Engineering Projects

Dust concentration in tunnel workplace   Table 1-4

No.	Name	Chemical abstracts (CAS No.)	PC-TWA/(mg/m^3) Total dust	PC-TWA/(mg/m^3) Whistling	Comments
1	Dust 10% ≤ free SiO$_2$ ≤ 50% 50% < free SiO$_2$ ≤ 80% free SiO$_2$ > 80%	148080-60-7	1 0.7 0.5	0.7 0.3 0.2	GI(Crystalline)

The occupational exposure limit (OEL) of chemical substances includes three indexes: time-weighted average permissible concentration (PC-TWA), permissible concentration of short-time exposure limit (PC-STEL) and maximum allowable concentration (MAC). The allowable concentration of dust only includes one index: time-weighted average permissible concentration (PC-TWA).

Time-weighted average permissible concentration (PC-TWA) is the average allowable exposure concentration (AEC) of 8 hrs working day and 40 hrs working week defined by time weight.

PC-STEL is the concentration at which short-time (15 mins) contact is allowed, subject to PC-TWA.

Maximum allowable concentration (MAC) is the concentration at the workplace at which toxic chemicals should not exceed at any time during a working day.

1.1.3 Foreign Standards

1. the United States

Occupational exposure limits have been recommended by the American Council of Industrial Hygienists and six academic societies. The Permissible Exposure Limits (PELs) for occupational hazards were published in the Federal Register by the Department of Labor's Occupational Safety and Health (OSHA). After public comment and amendment, PELs were promulgated in the 29th volume of the Federal Code as mandatory workplace health standards.

There are more than 650 mandatory occupational exposure limits issued by the Occupational Safety and Health Administration (OSHA) of the United States Department of Labor. Table 1-5 and Table 1-6 lists the mandatory occupational exposure limits related to the tunnel construction working environment.

Mandatory occupational exposure limits relevant to tunnel   Table 1-5
working environment in the United States

Chemical material	Chemical abstracts (CAS No.)	PELs[①] ppm[②]	PELs[①] (mg/m^3)[③]
Carbon dioxide	124-38-9	5 000	9 000
Carbon monoxide	630-08-0	50	55

continued

Chemical material	Chemical abstracts (CAS No.)	PELs[1]	
		ppm[2]	(mg/m^3)[3]
Hydrogen sulfide	7783-06-4	—	See Table 1-6
Nitrogen dioxide	10102-44-0	5[4]	9[4]
Sulfur dioxide	7446-09-5	5	13

Note: [1] PELs (unless otherwise noted) are 8 hrs TWAs.
[2] 10^{-6} concentration of volume of steam or gas at 25 ℃, 760 mmHg atmospheric pressure.
[3] The concentration unit is mg/m^3.
[4] indicates the upper limit.

Mandatory occupational exposure limits relevant to tunnel working environment in the United States　　Table 1-6

Chemical material	8 hrs weighted average	Acceptable upper limit concentration	8 hrs acceptable limit when the acceptable upper limit concentration is exceeded	
			concentration	Maximum duration
Hydrogen sulfide	—	20 ppm	50 ppm	10 mins, in case of last resort

2. Germany

Health standards for workplace chemicals issued by the German Federal Ministry of Labour and Social Affairs are divided into two parts: maximum allowable concentration (MAK) and biological tolerance value (BTA). MAK is usually the average of concentrations measured over a working day or shift rather than a single measurement. BTA is also based on a maximum daily exposure of 8 hours, 40 hours a week. In addition, the standard also makes clear the upper limit of exposure to chemical substances, the contact line is divided into short-term average and instantaneous value.

In 1996, there were 700 chemical substances in MAK and 44 chemical substances in BTA. Table 1-7 lists the chemicals MAK related to the tunnel construction working environment. Table 1-8 is the upper limit of contact with the chemical substances corresponding to Table 1-7.

Standards for chemical substances in German tunnel workplaces　　Table 1-7

Chemical material [CAS]	MAK		Exposure upper limits	H:S S(P)	Classification of carcinogens	Toxicity during pregnancy	Genotoxicity	Vapor pressure (hPa/20℃)
	(mL/m^3)	(mg/m^3)						
Carbon monoxide [630-8-0]	30	33	II.1			B		
Carbon dioxide [124-38-9]	5000	9000	IV					
Sulfur dioxide [7746-09-5]	2	5	I					
Hydrogen sulfide [7783-6-4]	10	14	V			II c		
Nitrogen dioxide [10102-44-0]	5	9	I					

Upper exposure limit of German chemical substances   Table 1-8

Type	Upper limit of exposure		Maximum number of exposures allowed per work shift
	MAK multiple	Duration	
I Local irritant	2	5 min, instantaneous value	8
II Systemic toxicity in 2 hrs			
II.1: Semi-reduction period <2 hrs	2	30 min, average value	4
II.2: Half-down period 2 hrs to 1 work shift	5	30 min, average value	2
III Systemic toxicity in 2 hrs			
Semi-reduction period >1 work shift (strong accumulation)	10	30 min, average value	1
IV Very weak substance			
MAK $> 500 \times 10^{-6}$	2	60 min, instantaneous value	3
V Substance with strong odor	2	10 min, instantaneous value	4

1.2 Natural Ventilation

Ventilation methods of underground engineering construction are divided into natural ventilation and mechanical ventilation according to power sources. Natural ventilation uses natural air pressure, while mechanical ventilation uses air pressure generated by ventilators. This section only discusses natural ventilation, mechanical ventilation is discussed in the next section.

Tunnel natural ventilation is a method to discharge pollutants from the tunnel without fan equipment and completely depend on the role of natural wind. It does not need equipment and electricity, but can saves energy and operation costs. It is an ideal ventilation method. But this method can not be used at will, it is restricted by the temperature difference inside and outside the tunnel, weather conditions, access adit setting, slope and other factors. To make use of natural ventilation, we need to understand its natural law.

1.2.1 Natural Air Flow in Tunnel

1. Formation of natural air flow

The formation of natural air flow in tunnel includes three reasons: temperature difference inside and outside the tunnel, transverse air pressure difference at the high point of air inlet and outlet, and natural air outside the tunnel.

1) Temperature difference

When the temperature inside and outside the tunnel is different, the density of the air inside and outside the tunnel is different. If there is a height difference between the air inlet and the air outlet, a pressure difference will be formed, which will result in the flow of air. This pressure difference is called thermal heterodyne. Of course, the temperature difference is not the only factor that causes the density difference, but the density difference is usually caused by the temperature difference.

2) Horizontal air pressure difference

In a large range, there are different climates in different places, air temperature and humidity are different, and the atmospheric pressure at the same level is also different, that is, there is a horizontal pressure difference. In meteorology, the pressure gradient is used to express the difference of the atmospheric pressure. The so-called pressure gradient is a vector perpendicular to the isobaric line, with 1 degree of a meridian or 111.1 km as a unit distance. The magnitude of the pressure change within each unit distance is called a pressure gradient.

It can be seen that the pressure gradient is a concept for larger ranges while the pressure difference at the same level in a small range is usually very small, can be ignored. However, in mountainous areas where you can experience all four seasons in one day, the temperature and humidity outside the entrance and exit of the extra-long tunnel are usually different, so the horizontal pressure difference cannot be ignored.

3) Atmospheric natural wind outside the tunnel

When natural wind blowing from outside toward the tunnel entrance hits the hillside, a part of its dynamic pressure can be converted into static pressure. The magnitude of this part of the power is related to the direction and speed of the natural wind in the atmosphere and is usually calculated as follows:

$$\Delta P_V = \frac{1}{2} \rho_a (v_a \cos\alpha)^2 \tag{1-1}$$

Where: ΔP_V——Atmospheric natural wind conversion dynamic pressure (Pa);

ρ_a——Atmospheric natural wind density outside tunnel (kg/m^3);

v_a——Natural atmospheric wind speed outside tunnel (m/s);

α——Angle between atmospheric natural wind direction and tunnel middle line (°).

2. Calculation of natural wind pressure

The natural air flow in the tunnel is caused by the temperature difference inside and outside the tunnel, the horizontal air pressure difference at the entrance and exit high point and the natural air flow outside the tunnel. The air pressure of the natural air flow is the sum of the three factors. When calculating, the reference point can be either the low hole or the high hole. As shown in Figure 1-1, it is the natural ventilation after the tunnel entrance work area is connected with the vertical shaft work area. 1 point is the air inlet of the tunnel, and 3 point are the air outlet of the tunnel. In other words, the natural air flow enters from one point, passes through two points, and is finally discharged from three points. The average temperature outside the tunnel is T_1, the average temperature outside the tunnel is T_2, the atmospheric pressure at 1 point is P_1, the atmospheric pressure at 3 points is P_3, 0-3 is the high-point horizontal line, the atmospheric pressure at 0 point above 1 point is P_0, and the angle between the natural wind direction outside the tunnel entrance at 1 point and the tunnel center line is, taking 1 point as the reference point,

the natural wind pressure is:
$$H_N = \Delta P_V + (P_3 - P_0) + (\rho_{m1} - \rho_{m2})g \cdot Z \tag{1-2}$$
Where: ρ_{m1} ——Average density of air at points 0-1;

ρ_{m2} ——Average density of air at points 2-3.

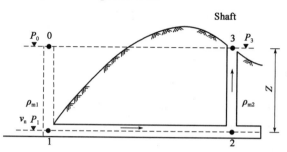

Figure 1-1 Pressure relation forming natural air flow in tunnels

The first term on the right is the dynamic pressure of the part where the atmospheric air flow outside the tunnel inlet is converted to the static pressure, and the value can be calculated according to Formula (1-1); The second term is the horizontal pressure difference between high point 3 and zero point; The third term represents the thermal heterodyne due to the temperature difference between the inside and outside of the tunnel, that is, the gravity difference between the two sides of the air column.

3. Influencing factors of natural wind pressure

Usually, the change of dynamic pressure of atmospheric air flow outside the tunnel and the horizontal air pressure at high point have little influence on the natural air pressure, and the magnitude of the natural air pressure mainly depends on the temperature head. The air density is not only affected by temperature T, but also by atmospheric pressure P, relative humidity φ and partial pressure of saturated water vapor in the air. The main results are listed as follows:

(1) The temperature difference between the two sides of the air column is the main factor affecting the natural wind pressure. The main factors affecting the temperature difference are the temperature outside the tunnel, the air flow into the tunnel, the surrounding rock temperature and the heat exchange between the air flow and the surrounding rock. The degree of influence varies with the tunnel construction method, buried depth, topography, season and geographical location.

(2) The air composition and humidity affect the density of air and the natural wind pressure to a certain extent, *but the influence is small.*

(3) When the air density difference between the two sides is constant, the temperature head is proportional to the height difference Z between the highest and lowest (horizontal) points. *That is, the greater the height difference, the greater the temperature head.*

4. Calculation of natural air volume

In natural ventilation, the energy is expressed as natural ventilation pressure H_N, which can be calculated according to Formula (1-2).

The natural air volume Q_N is determined by the natural ventilation pressure and ventilation resistance. When the pressure loss $h = H_N$ in the tunnel, it can be calculated by the following formula:

$$Q_N = 60 \cdot A \cdot v \tag{1-3a}$$

$$v = \sqrt{H_N \cdot \frac{1}{\lambda} \cdot \frac{2}{\rho} \cdot \frac{d_T}{L_T}} \tag{1-3b}$$

Where: Q_N——Natural ventilation (m³/min);
       A——Sectional area of tunnel (m²);
       v——Average velocity in tunnel (m/s);
       λ——Friction coefficient of tunnel inner wall;
       L_T——Tunnel length (m);
       d_T——Tunnel equivalent diameter (m);
       ρ——Air density in tunnel (kg/m³).

1.2.2 Natural Ventilation of Tunnels in Common Situations

1. Natural ventilation for single head construction of uphill and downhill tunnels

The formation of natural ventilation is closely related to the climatic conditions at the entrance of the tunnel when the entrance and exit of the tunnel are constructed on the uphill and downhill slopes.

For downhill tunnels, see Figure 1-2. In winter, generally speaking, the temperature inside the tunnel is higher than that outside the tunnel, and the cold air outside will enter the tunnel along the lower part of the tunnel, and the warm air containing pollutants inside the tunnel will be discharged out of the tunnel along the upper part of the tunnel to form natural wind; In summer, the temperature inside the tunnel is lower than that outside the tunnel. The hot air outside the tunnel is blocked at the entrance of the tunnel, and the cool air inside the tunnel is stagnant outside the tunnel, which can not form natural air flow.

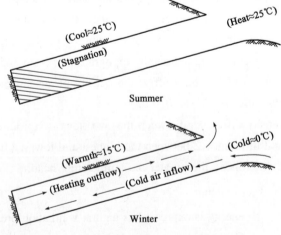

Figure 1-2 Natural ventilation schematics for downhill tunnels

For tunnels constructed on the uphill slope, as shown in Figure 1-3, in winter, the cold air outside the tunnel stops at the entrance of the tunnel, and it is difficult to enter the working face. The hot air containing pollutants in the tunnel is blocked near the working face by the natural wind pressure, so it is difficult to form the natural wind flow. In summer, the temperature inside the tunnel is low, the cool air inside the tunnel flows to the lower part, and the hot air outside the tunnel flows in from the upper part of the tunnel entrance, which results in natural ventilation.

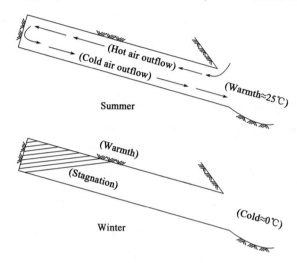

Figure 1-3 Natural ventilation schematics for uphill tunnels

2. Natural ventilation after two vertical (inclined) shaft working areas connected

When the two vertical (inclined) shafts areas are connected, natural ventilation is easy to form between them because of the different position and depth of the two vertical (inclined) shafts. The natural ventilation system is shown in Figure 1-4. Line 2-3 is the horizontal tunnel, and Line 0-4 is the horizontal line at the highest point of the ventilation system. If the surface atmosphere is regarded as an imaginary air path with infinite cross-section and zero wind resistance, the ventilation system is regarded as a closed loop.

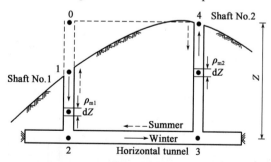

Figure 1-4 Natural ventilation schematics of tunnel after two shaft working areas connected

If the natural wind outside the tunnel and the horizontal pressure difference between 4 and 0 points are not taken into account. The magnitude and direction of natural wind pressure are mainly affected by the change of surface air temperature. In winter, the surface temperature is very low, the air column 0-1-2 is heavier than the air column 4-3, the air flow flows from the vertical shaft No. 1 to the vertical shaft No. 2, and then flows to the surface through the vertical shaft No. 3 and vertical shaft No. 4. In summer, the surface temperature is higher than the average temperature in the tunnel and shaft, and the air column 0-1-2 is lighter than the air column 4-3, so that the air flow is discharged from the shaft No. 1. In spring and autumn, the heading face temperature is not much different from the average temperature in the tunnel shaft, and the natural wind pressure is small, so the wind flow in the tunnel is stagnant. In some mountainous areas, the day and night temperatures on the ground vary greatly, so the natural wind pressure will also change accordingly. At night, that air enter the vertical shaft No. 1; At midday, the wind blew out of the vertical shaft No. 1.

3. Natural ventilation in tunnels with vertical shaft

When blind heading in a long tunnel is under construction, if the terrain conditions are available, ventilation shaft will be set up to reduce the ventilation length of blind heading and the ventilation difficulty, meanwhile, natural wind can be fully utilized and energy consumption can be reduced in the season when the temperature difference between the inside and outside of the tunnel is large.

Figure 1-5 are ventilation diagrams of a tunnel with a ventilation shaft. In this case, Point 0 and Point 3 on the same side of the hillside. The distance between Point 0 and Point 3 is close, the influences of horizontal atmospheric pressure difference and natural wind outside the tunnel are weak and ignored. The magnitude of the natural wind pressure depends mainly on the density difference of air columns 0-1 and 2-3 due to the temperature difference between the inside and outside of the tunnel, and the depth of the ventilation shaft. In other words, the natural wind pressure can be calculated according to Formula (1-2). The magnitude of the air volume depends on the magnitude of the wind resistance of the ventilation path 1-2-3. In winter, the outside temperature is lower than that in the tunnel, the density of air column $0-1$ is higher than that of 2-3, and the natural air flow in the tunnel enters from 1 point and exits from 3 point. In summer, the opposite is true.

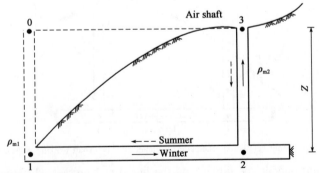

Figure 1-5 Natural ventilation schematics for tunnels with vertical shaft

1.3 Basic Mechanical Ventilation

1. Supply ventilation

The air inlet of the supply ventilation duct is arranged outside the tunnel, and the air outlet is arranged near the heading face. Under the action of the fan, fresh air is sent to the heading face from outside the tunnel through the duct to dilute pollutants and polluted air is discharged from the tunnel to outside the tunnel, as shown in Figure 1-6.

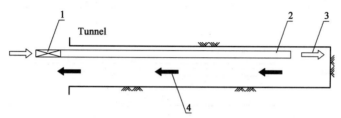

Figure 1-6 Supply ventilation schematics
1-fan;2-exhaust air duct;3-fresh air;4-polluted air

2. Exhaust ventilation

Exhaust ventilation is divided into negative pressure exhaust ventilation and positive pressure exhaust ventilation. The air inlet of the duct is arranged near the heading surface, and the air outlet is arranged outside the tunnel. Under the action of the fan, fresh air reaches the heading surface from outside the tunnel through the tunnel, and polluted air is discharged directly from the duct to outside the tunnel, as shown in Figure 1-7 and Figure 1-8.

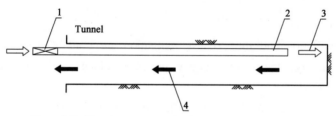

Figure 1-7 Negative pressure and exhaust ventilation schematics
1-fan;2-exhaust duct;3-polluted air;4-fresh air

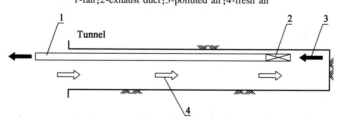

Figure 1-8 Positive pressure and exhaust ventilation schematics
1-exhaust duct;2-fan;3-polluted air;4-fresh air

3. Compound ventilation

Thecompoundventilation is a combination of supply ventilation and exhaust ventilation transformations. It has two forms, one is negative pressure exhaust compound type, the other is positive pressure exhaust compound type, as shown in Figures 1-9 and 1-10.

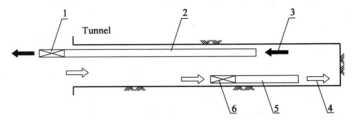

Figure 1-9 Diagram of negative pressure and exhaust compound ventilation
1-fan;2-exhaust duct;3-polluted air;4-fresh air;5-supply duct;6-air blower

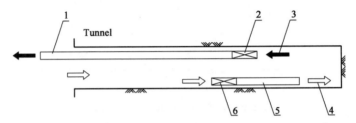

Figure 1-10 Diagram of positive pressure and exhaust compound ventilation
1-exhaust duct;2-exhaust fan;3-polluted air;4-fresh air;5-supply duct;6-air blower

Fresh air enters the tunnel from outside the tunnel, flows to the inlet of the fan and enters the supply duct driven by fan blades, The fresh air is sent to the heading face through the supply duct; From the heading face of the tunnel, the polluted air passes through the inlet of the exhaust duct, enters the exhaust duct, and is discharged to the outside of the tunnel through the exhaust duct.

4. Combined type

The supply ventilation and the exhaust ventilation are used at the same time to form a combined type. Similarly, there are negative pressure exhaust and positive pressure exhaust combined with two forms.

A part of fresh air is sent to the heading face through the supply duct, A part of fresh air enters the tunnel through the tunnel from the outside of the tunnel, and a part of polluted air flows from the heading face to the inlet of the supply duct; the other part of fresh air enters the tunnel along the way to dilute the dirty matter into polluted air and then flows to the inlet of the exhaust duct; two streams of polluted air flow into the exhaust duct and are discharged out of the tunnel, as shown in Figure 1-11 and Figure 1-12, respectively.

5. Gallery ventilation

The gallery ventilation is divided into jet gallery type and main fan gallery type. Jet fans drive fresh air into one tunnel and out from the other side; it is call the jet gallery type. Fresh air is delivered to the heading face through the supply duct. The system arrangement is shown in Figure 1-13.

Chapter 1 Construction Ventilation of Underground Engineering Projects

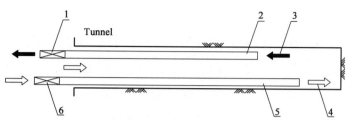

Figure 1-11 Schematics of negative pressure and exhaust combined ventilation
1-exhaust fan;2-exhaust duct;3-polluted air;4-fresh air;5-supply duct;6-air blower

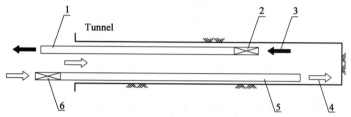

Figure 1-12 Schematics of positive pressure and exhaust combined ventilation
1-exhaust fan;2-exhaust duct;3-polluted air;4-fresh air;5-supply duct;6-air blower

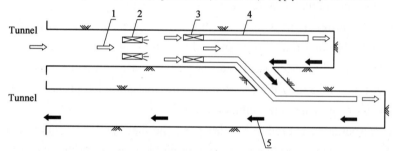

Figure 1-13 Schematic diagram of jet gallery ventilation
1-fresh air;2-jet fan;3-air blower;4-supply duct;5-polluted air

The main fan gallery type is under the main fan function, the fresh air enters from one tunnel, and the polluted air is discharged from the other tunnel. Fresh air is delivered to the heading face through the supply duct. The system arrangement is shown in Figure 1-14.

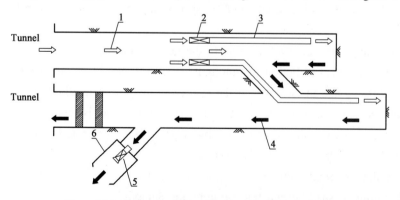

Figure 1-14 Schematic diagram of main fan gallery ventilation
1-fresh air;2-air blower;3-supply duct;4-polluted air;5-main fan;6-damper

1.4 Ventilation Methods in Common Tunnels with Access Adits

In a one-tube tunnel without access adit, the inlet and outlet are constructed by blind heading, while the entrance and exit of a twin-tube tunnel are constructed by parallel tunneling, and the ventilation is relatively simple. In order to increase the working face, shorten the construction period and improve the construction conditions, it is necessary to add access adit. The common access adits are transverse passage, parallel adit, inclined shaft and vertical shaft. Due to the setting of access adits, the ventilation should be changed correspondingly.

1.4.1 Ventilation Method of Single-hole Tunnels

1. Ventilation for blind heading construction at the inlet (or outlet)

Borehole-blasting method is adopted in the construction of the tunnel, and trackless transportation or rail transportation is selected.

When trackless transportation is used in construction, there are two main types of pollution sources: one includes blasting fume, dust and tail gas from internal combustion and slag loading equipment, which is mainly concentrated near the heading face and is a relatively fixed type; the other is the exhaust gas emitted by diesel vehicles in the transportation projects, which pollutes the whole tunnel and isa type of moving pollution source.

When rail transportation is used in construction, the main pollution source is blasting fume, which is concentrated near the heading face. In transportation engineering, battery cars or electric locomotives do not produce exhaust fumes.

The method of transportation usually affects the ventilation choice.

1) For tunnels constructed with trackless transportation

(1) Ventilation method

Usually, supply ventilation is used, and the system arrangement is shown in Figure 1-6.

(2) Features

①Fresh air can be delivered all the way to the heading face.

②The concentration distribution of automobile exhaust gas in the tunnel tends to increase from inside to outside, and the workers in the operation area are in relatively fresh air.

③The soft air duct can be used, and the extension of the duct is relatively easy.

④The whole tunnel is polluted and the subsequent working environment is relatively poor.

⑤Leakage of ducts has positive effect on ventilation.

(3) Key points for implementation

①Ventilation ducts shall be laid flat, straight and smooth.

②The distance from the air outlet to the heading face is less than 5 times of the equivalent diameter of the tunnel.

③The distance between the fan and the tunnel entrance is about 10 times of the equivalent diameter of the tunnel, or one side of the tunnel entrance is arranged in a right angle and kept at a certain distance.

2) Construction of rail tunnel transportation

(1) Ventilation method

When the blind heading distance is short, supply ventilation is adopted, and its system arrangement and implementation points are the same as those of the trackless transport tunnel, and its characteristics are the same as those of trackless transport tunnel except for no automobile exhaust.

When the blind heading distance is long, the negative pressure exhaust mixed type or the positive pressure exhaust mixed type can be used, and the arrangement is shown in Figures 1-9 and 1-10.

(2) Features

①Fresh air is transferred to the heading face.

②The area from the air inlet of the exhaust ducts to the entrance of the tunnel is in the fresh air.

③Soft air duct can be used and easy to be extended.

④Soft air duct can be used in positive pressure exhaust ducts, but it is not easy to extend, and exhaust fan must be moved at the same time; The negative pressure exhaust ducts can not use the soft air duct, so the cost is high.

⑤The air leakage of positive pressure exhaust ducts has negative effect on ventilation and will cause secondary pollution, while the air leakage of negative pressure exhaust ducts does not cause secondary pollution.

⑥Noise pollution in tunnel is easily formed by blower and positive pressure exhaust blower.

(3) Key points for implementation

①Both the blower and the exhaust fan are arranged on the side of the tunnel opening near the lining formwork. With the advancement of the heading face, the supply duct follows the heading face closely.

②Ventilation ducts shall be laid flat, straight and smooth.

③The distance between the air outlet of the supply duct and the heading face is less than 5 times of the equivalent diameter of the tunnel.

④The distance between the air outlet of the exhaust ducts and the opening of the tunnel is about 10 times of the equivalent diameter of the tunnel or a certain angle upward in the direction of right angle.

⑤The overlapping length of the supply duct and the air exhaust duct is about 50 m.

⑥The noise of the blower and positive pressure exhaust fan must meet the standard requirements.

2. Ventilation method for blind heading construction at the inlet (or outlet) with ventilation shaft

1) For tunnels constructed with trackless transportation

(1) Ventilation method

When the natural air flows from the tunnel entrance to the ventilation shaft, the jet gallery ventilation shown in Figure 1-15 can be used. Fresh air is directly sent to the heading face by the supply duct, and the polluted air on the heading face flows to the ventilation shaft along the way, and is discharged out of the tunnel through the shaft. The air flow from the opening to the ventilation shaft can be adjusted by jet fan injection. If the natural air flow is large enough, the jet fan can be turned off.

When the natural air flows from the ventilation shaft to the tunnel opening, or the distance from the heading face to the ventilation shaft is too long, the supply ventilation shown in Figure 1-16 may be used. Fresh air is directly sent to the heading face of the tunnel through the supply duct from the ventilation vertical shaft, and the polluted air flows from the heading face to the tunnel opening and is discharged to the outside of the tunnel.

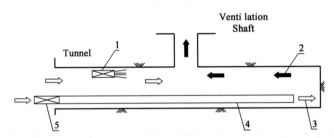

Figure 1-15 Diagram of jet gallery ventilation for blind heading in single-hole tunnel with ventilation shaft
1-jet fan;2-polluted air;3-fresh air;4-supply duct;5-air blower

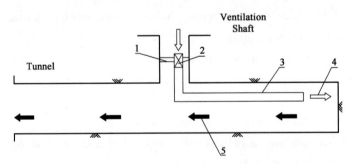

Figure 1-16 Diagram of supply ventilation for blind heading in single-hole tunnel with ventilation shaft
1-air barrier;2-air blower;3-supply duct;4-fresh air;5-polluted air

(2) Features

①Fresh air can be delivered all the way to the heading face.

②When adopting jet gallery ventilation as shown in Figure 1-15, after balancing, the concentration distribution of automobile exhaust gas in the tunnel increases gradually from the opening to the vertical shaft and from the heading face to the vertical shaft, and the concentration in the vertical shaft is the highest. When the supply ventilation as shown in Figure 1-16 is adopted, after balancing, the concentration distribution of automobile exhaust gas in the tunnel increases gradually from the inside to the outside, and the concentration at the opening of the tunnel is the highest. Working area workers in both ventilation systems are exposed to relatively fresh air.

③Soft air duct can be used, and the extension of the supply duct is relatively easy.

④The whole tunnel is polluted and the subsequent working environment is relatively poor.

⑤Leakage of duct has positive effect on ventilation.

(3) Key points for implementation

①The distribution of supply ducts shall be flat, straight and smooth.

②The distance from the air outlet to the heading face is less than 5 times of the equivalent diameter of the tunnel.

③When jet gallery ventilation is adopted, the distance between the blower and the tunnel opening is about 10 times of the equivalent diameter of the tunnel or a certain angle upward in the direction of right angle.

2) For tunnels constructed by rail transportation

(1) Ventilation method

When the natural air flows from the tunnel opening to the ventilation vertical shaft, the jet gallery ventilation shown in Figure 1-17 can be used. Under the action of jet fan and natural wind, when the fresh air flow reaches the inlet of the blower, the fresh air flow is sent to the heading face of the blower through the supply duct, and the polluted air flows from the heading face to the ventilation vertical shaft, and the polluted air is discharged from the vertical shaft to the outside of the tunnel. If the natural air flow is large enough, turn off the jet fan.

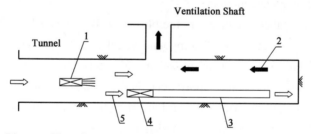

Figure 1-17 Diagram of Jet gallery ventilation for blind heading construction with ventilation shaft
1-jet fan; 2-polluted air; 3-supply duct; 4-air blower; 5-fresh air

When the natural air flows from the ventilation vertical shaft to the tunnel opening or the distance from the heading face to the ventilation vertical shaft is too long, the positive pressure exhaust compound ventilation shown in Figure 1-18 may be used. Fresh air flows from the outside of the tunnel to the inside of the tunnel. When the fresh air reaches the inlet of the blower, the fresh air is sent to the heading face by the supply duct. The polluted air flows from the heading face to the inlet of the blower of the exhaust duct, enters the exhaust duct, and is discharged out of the tunnel through the ventilation vertical shaft.

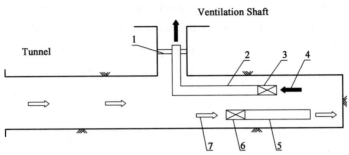

Figure 1-18 Diagram of positive pressure and exhaust compound ventilation for blind heading Construction with Ventilation Shaft
1-air baffle;2-exhaust duct;3-air blower;4-polluted air;5-supply duct;6-air blower;7-fresh air

(2) Features

①Fresh air is transferred to the heading face.

②When jet gallery ventilation is adopted, the area from the vertical shaft to the tunnel opening is in fresh air; when compound ventilation is adopted, the area from the exhaust fan inlet to the tunnel opening is in fresh air.

③Soft air duct can be used in the supply duct, and the extension of the duct is relatively easy.

④Soft air ducts can be used in the exhaust duct, but it is difficult to extend, and the exhaust duct must be moved at the same time.

⑤Air leakage from exhaust duct has negative effect on ventilation and will cause secondary pollution.

⑥Large ventilation section and low power consumption.

⑦The air duct is little and the cost is low.

⑧Noise pollution in the tunnel is easy to be caused by the fan in the tunnel.

(3) Key points for implementation

①In the case of jet gallery ventilation, the blower shall be located on the upwind side of the ventilation vertical shaft. In the case of compound ventilation, both the blower and the exhaust blower shall be located on the tunnel opening side near the lining formwork. With the advancement of the heading face, the supply duct shall follow the heading face closely.

②Ventilation ducts shall be arranged in a flat, straight and smooth manner, especially at the corners where the exhaust ducts enter the ventilation vertical shaft, so as to be smooth without turning into a dead angle.

③The distance between the air outlet of the supply duct and the heading face is less than 5 times of the equivalent diameter of the tunnel.

④The overlapping length of mixed feeding and discharging ducts are less than 50 m.

⑤The fan in the tunnel must meet the requirements of noise standard.

3. Ventilation from transverse gallery to tunnel

1) Ventilation method

The jet gallery ventilation shown in Figure 1-19 may be used when entering the tunnel from the transverse gallery for construction and penetrating the tunnel opening. Under the action of the jet fan, the new air flows from the tunnel to the transverse gallery. When it reaches the inlet of the blower, the fresh air is sent to the heading face by the supply duct. The polluted air flows from the heading face to the transverse gallery and is discharged from the transverse gallery.

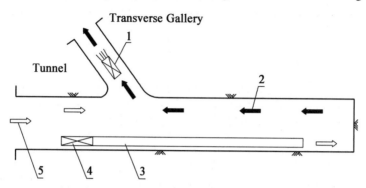

Figure 1-19　Diagram of jet gallery ventilation for construction with transverse gallery entering into tunnel
1-fresh air;2-air blower;3-supply duct;4-polluted air;5-jet fan

2) Features

①Fresh air is transferred to the heading face.

②The area from the tunnel opening to the transverse gallery is fresh air flow.

③Soft air duct can be used in the supply duct, and the extension of the duct is relatively easy.

④The supply distance of duct is short, and the air duct needs less.

⑤Leakage of duct has positive effect on ventilation.

3) Key Points for Implementation

①The jet fan is preferably installed in the transverse gallery.

②The ventilation ducts shall be laid flat, straight and smooth.

③The distance between the outlet of the vent duct and the heading face of the tunnel is less than 5 times of the equivalent diameter of the tunnel.

④The distance from the outlet of the blower to the transverse gallery shall be more than 50m.

4. Ventilation method for tunnel with parallel adit

Parallel tunneling is conducted forward in the tunnel and its parallel adit. The tunneling in the parallel adit goes faster in order to create more work faces for the tunnel through the cross channel between the tunnel and its parallel adit.

1) Ventilation method

When the blind heading distance is short, the tunnel with parallel adit adopt the supply ventilation, and the system layout and implementation points are the same as those of the blind heading construction of single-hole tunnel.

When the blind heading distance is long, jet gallery ventilation is adopted. The arrangement of jet gallery ventilation system is shown in Figure 1-20. Jet gallery ventilation uses jet fan to form air flow in a tunnel with parallel adit. Fresh air is introduced into the tunnel through one hole, and when the fresh air flows near the unblocked cross channel, the fresh air upstream of the cross channel is sent to the heading faces of the parallel adit and the two tunnels through three ducts, and the polluted air flows back to the cross channel from the heading face and is discharged from the other hole.

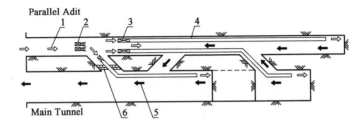

Figure 1-20 Diagram of jet gallery ventilation for parallel construction of a tunnel with parallel adit
1-fresh air;2-jet fan;3-supply fan;4-supply duct;5-polluted air;6-wind wall

2) Features

①Fresh air is transferred to the heading face.

②The fresh air flow is from the inlet of the air intake tunnel to the area of the unblocked cross channel.

③Soft air duct can be used in the supply duct, and the extension of the duct is relatively easy.

④Large ventilation section and low power consumption.

⑤Few air ducts are needed and the cost is low.

3) Key points for implementation

(1) The jet fan is preferably installed in the parallel adit.

(2) Unused cross channel shall be closed in time. For construction purpose, air doors shall be installed for those that cannot be closed. If air doors cannot be installed, jet fans shall be used for regulation and control.

(3) Ventilation ducts shall be laid flat, straight and smooth, especially the ventilation ducts entering another tunnel from the cross channel of the intake tunnel at the corner of the cross channel without turning dead angle.

(4) The distance between the air outlet of the supply duct and the heading face of the parallel adit (or tunnel) is less than 5 times the equivalent diameter of the parallel adit (or tunnel).

(5) When trackless transportation is used for construction, vehicles must enter through the exhaust hole.

5. Ventilation method for simultaneous construction of multiple working faces with parallel adit

1) Ventilation method

The jet gallery ventilation is usually adopted, and the system arrangement is shown in Figure 1-21.

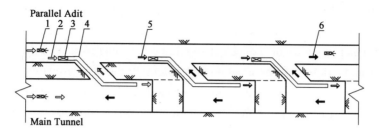

Figure 1-21 Schematic diagram of Jet Gallery ventilation for simultaneous construction of multiple working faces through a parallel adit (the blowers are placed separately)

1-jet fan;2-fresh air;3-air blower;4-supply duct;5-lightly polluted air;6-polluted air

A large enough main air flow is in that through parallel adit by using natural air and jet fan. The blowers are arranged separately on the windward side of each cross channel. The supply duct enters into different operation areas through each cross channel, and the air outlet is located near the heading face. Fresh air is sent from the parallel adit to the heading face through the supply duct, diluting pollutants, and polluted air enters the cross channel through the tunnel, enters the main air flow of the parallel adit, and is discharged out of the tunnel along the air flow.

In special cases, such as trackless transportation, the centralized placement of blowers can be adopted, as shown in Figure 1-22.

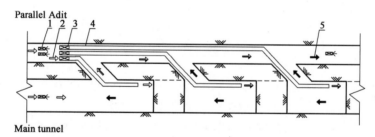

Figure 1-22 Schematic diagram of jet gallery ventilation for simultaneous construction of multiple working faces through a parallel adit (concentrated placement of blowers)

1-jet fan;2-fresh air;3-air blower;4-supply duct;5-polluted air

2) Features

(1) Fresh air is transferred to the heading face.

(2) Soft air duct can be used in the supply duct, and the extension of the duct is relatively easy.

(3) Large ventilation section and low power consumption.

(4) Duct air leakage has positive effect on ventilation.

(5) When the blowers are placed separately, the blowers provided in the downdraft flow are not completely fresh air delivered to the working surface.

3) Key points for implementation

(1) The distribution of supply ducts shall be flat, straight and smooth, especially at corners without turning dead corners.

(2) The distance between the air outlet of the supply duct and the heading face is less than 5 times of the equivalent diameter of the tunnel.

(3) The ejecting direction of the jet fan should be the same as that of the natural wind.

(4) The main air flow shall be large enough to ensure the supply quality of the blower.

(5) The blower must be arranged on the windward side of the cross channel.

(6) When the upstream blasting fume passes by, turn off the downstream blower temporarily.

(7) During trackless transportation, vehicles enter through the air outlet.

6. Ventilation method for two-way construction from inclined shaft into tunnel

The two-way construction from inclined shaft into tunnel means blind heading in two directions after entering the tunnel through inclined shaft, that is single inclined shaft and single main tunnel method.

1) For tunnels constructed with trackless transportation

(1) Ventilation method

Both work surfaces are ventilated by supply, and the system layout is shown in Figure 1-23.

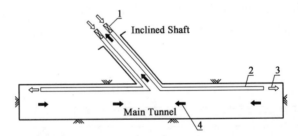

Figure 1-23 Schematic diagram of supply ventilation for two-way construction of single-hole tunnel with single inclined shaft
1-fan;2-supply duct;3-fresh air;4-pollution air

Both fans are located outside the tunnel, and the air outlets of the two supply ducts are located near the two heading faces respectively. Fresh air is sent from the outside of the tunnel to the heading face through the supply duct, and polluted air flows from the two heading faces to the inclined shaft, and then is discharged out of the tunnel through the inclined shaft.

(2) Features

①Fresh air can be delivered all the way to the heading face.

②After balancing, the concentration distribution of automobile exhaust gas in the tunnel increases gradually from inside to outside, the concentration near the inclined shaft opening is the highest, and the workers in the operation area are in relatively fresh air.

③The soft air duct can be used, and the extension of the duct is relatively easy.

④The whole tunnel and inclined shaft were polluted, and the subsequent working environment was relatively poor.

⑤Leakage of duct has positive effect on ventilation.

(3) Key points for implementation

①Ventilation ducts shall be arranged in a flat, straight and smooth manner, especially when the ducts are transferred from the inclined shaft to the main tunnel, and shall not be turned into a dead angle.

②The distance from the air outlet to the heading face is less than 5 times of the equivalent diameter of the tunnel.

③The distance between the blower and the tunnel opening is about 10 times of the equivalent diameter of the inclined shaft or placed on one side of the inclined shaft opening in a right-angle direction and kept at a certain distance.

2) For tunnels constructed by rail transportation

(1) Ventilation method

When the blind heading distance is short, supply ventilation shall be adopted. The system arrangement and implementation points are the same as those described above.

When the blind heading distance is long, the compound ventilation is adopted. The system arrangement is shown in Figure 1-24.

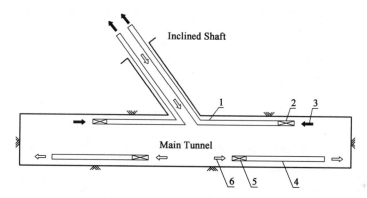

Figure 1-24 Schematic diagram of compound ventilation for two-way construction of single-hole tunnel with single inclined shaft

1-exhaust duct;2-exhaust fan;3-polluted air;4-supply duct;5-air blower;6-fresh air

For the mixed type, fresh air enters the main tunnel from the outside of the tunnel through the inclined shaft, then flows in two directions respectively to the inlet of the blower, and then passes through the supply duct to the heading face, and the polluted air flows from the heading face to the inlet of the exhaust fan, and then discharges out of the tunnel through the exhaust duct.

(2) Features

①Fresh air is transferred to the heading face.

②The area from the inlet of the exhaust duct to the opening of the inclined shaft is in fresh air.

③Soft air duct can be used in the supply duct, and the extension of the duct is relatively easy.

④Soft air duct can be used in the exhaust duct, but it is not easy to extend. The exhaust fan must be moved at the same time.

⑤Air leakage from exhaust duct has negative effect on ventilation and will cause secondary pollution.

⑥Noise pollution in tunnel is easy to be caused by fans in tunnel.

(3) Key points for implementation

①Both the blowers and the exhaust fans are arranged at the inclined shaft side behind the lining formwork, and the supply ducts shall be extended one by one to follow up with the advancement of the heading face.

②Ventilation ducts shall be laid flat, straight and smooth, especially when the ducts are transferred from the inclined shaft to the main tunnel, and shall not be turned into a dead angle.

③The distance between the air outlet of the supply duct and the heading face is less than 5 times of the equivalent diameter of the tunnel.

④The air outlet of the exhaust duct shall be about 10 times the equivalent diameter of the inclined shaft away from the tunnel opening, or shall be positioned above the tunnel opening in a right-angle direction.

⑤The overlapping length of the supply duct and the air exhaust duct shall not be less than 50 m.

⑥The fan in the tunnel must meet the noise requirements in the tunnel.

7. Ventilation method for two-way tunneling from vertical shaft to tunnel

Two-way tunneling from vertical shaft to tunnel refers to the tunneling after the vertical shaft enters two directions. Through the vertical shaft into the main tunnel construction, the city uses rail transportation method, so only consider rail transportation construction, that is, single inclined shaft and the main tunnel method.

1) Ventilation method

When the blind heading distance is short, supply ventilation is adopted, and the system arrangement and implementation points are the same as those of supply ventilation.

When the blind heading distance is long, the compound ventilation is adopted. The system arrangement is shown in Figure 1-25.

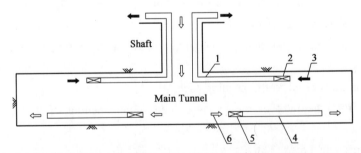

Figure 1-25 Schematic diagram of compound ventilation for two-way construction of single-hole tunnel with single shaft
1-exhaust duct;2-exhaust fan;3-polluted air;4-supply duct;5-air blower;6-fresh air

For the mixed type of supply and discharge, fresh air enters the main tunnel from outside the tunnel through the vertical shaft, then flows in two directions respectively to the inlet of the blower, and then is sent to the heading face through the supply duct; The polluted air flows from the tunnel to the inlet of the exhaust fan from the heading face, and then is discharged out of the tunnel through the exhaust duct.

2) Features

(1) Fresh air is transferred to the heading face.

(2) Fresh air flows from the inlet of the exhaust duct to the bottom hole area of the vertical shaft and the vertical shaft body of the shaft.

(3) Soft air duct can be used in the supply duct, and the extension of the duct is relatively easy.

(4) Soft air ducts can be used in the exhaust duct, but the extension is not easy, and the exhaust fan must be moved at the same time.

(5) Air leakage from exhaust duct has negative effect on ventilation and will cause secondary pollution.

(6) Noise pollution in tunnel caused by fan is easy to occur.

3) Key points for implementation

(1) Set both the blower and exhaust fan on the side of the vertical shaft behind the lining formwork. With the advancement of the heading face, the blower ducts shall be followed up one by one.

(2) Ventilation ducts shall be arranged in a flat, straight and smooth manner, especially at the position where the ducts are turned from the bottom of the vertical shaft to the main hole, so as to be smooth but not turn into a dead angle.

(3) The distance between the air outlet of the exhaust duct and the heading face is less than 5 times of the equivalent diameter of the tunnel.

(4) The diameter of the air duct about 10 times the transverse distance of the air outlet of the exhaust duct from the tunnel opening.

(5) The overlapping length of the supply duct and the air exhaust duct shall not be less than 50 m.

(6) The fan must meet the noise standard requirements.

1.4.2 Ventilation Method of Double-hole Tunnel

1. Ventilation method for parallel construction of inlet (or outlet)

The two work faces forward parallel tunneling, and the two tunnels are connected by a transverse gallery, that is, the parallel double-hole method.

Ventilation method, characteristics and key points of implementation are basically the same as that of parallel construction of tunnels with parallel adit. The difference is that the working surface is seldom added, and the arrangement of the jet gallery ventilation system is shown in Figure 1-13.

2. Ventilation method for parallel construction from cross channel into tunnel

Concurrent construction of double-hole tunnel from cross channel means that after entering parallel double-hole through cross channel, excavation is started in two directions and four working surfaces are constructed at the same time, but the direction to the opening of the tunnel is usually short, which will be penetrated quickly and become single-direction parallel construction of double-hole. The two holes are connected by transverse galleries at regular intervals.

Its ventilation method, characteristics and key points of implementation are approximately the same as that of the jet gallery ventilation in the parallel construction of the entrance and exit of the double-hole. The difference is:

(1) In the initial stage of the blind heading supply method, the ventilation ducts enter the tunnel through the transverse gallery.

(2) In the stage of jet gallery ventilation, if rail transportation is adopted, the blower shall be arranged in the tunnel directly connected with the transverse gallery, the fresh air flow shall enter the tunnel connected with the transverse gallery, and the polluted air shall be discharged from another tunnel, as shown in Figure 1-26. If trackless transportation is used, the blower shall be arranged in a tunnel not connected to the transverse gallery through which fresh air flows in and polluted air is discharged from the tunnel connected to the transverse gallery, as shown in Figure 1-27.

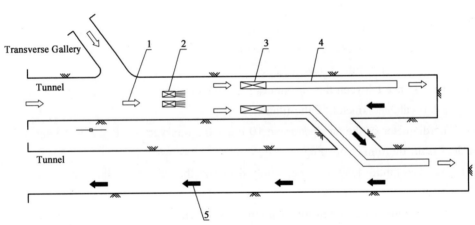

Figure 1-26 Schematic diagram of jet gallery ventilation for parallel construction with transverse gallery into double-hole tunnel (track transportation)

1-fresh air; 2-jet fan; 3-air blower; 4-supply duct; 5-polluted air

3. Two-way parallel construction method from the inclined shaft to the tunnel

Concurrent construction from inclined shaft into double-hole means that after entering parallel double-hole through inclined shaft, excavating in two directions, four working surfaces are constructed at the same time, and there is a cross channel connecting the two tunnels with a certain distance between them, which is called double tunnel method of single inclined shaft.

The transportation method is usually rail transportation, so the construction of rail transportation is considered here.

Chapter 1 Construction Ventilation of Underground Engineering Projects

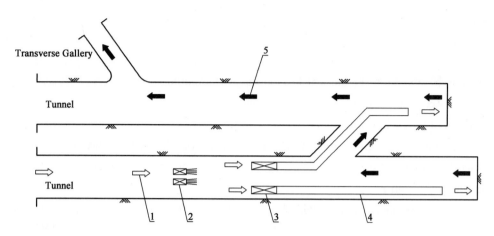

Figure 1-27 Schematic diagram of jet gallery ventilation for parallel construction with transverse gallery into double-hole tunnel (trackless transportation)

1-fresh air; 2-jet fan; 3-air blower; 4-supply duct; 5-polluted air

1) Ventilation method

When the blind heading distance is short, supply ventilation is adopted, and the system arrangement and key points of implementation are approximately the same as those of blind heading construction.

When the blind heading distance is long, the combination of positive pressure and exhaust ventilation and jet gallery ventilation is adopted. The system arrangement is shown in Figure 1-28.

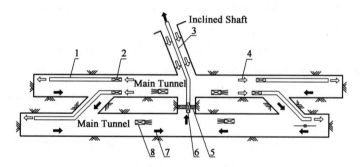

Figure 1-28 Schematic diagram of the ventilation system in a double-hole tunnel with single inclined shaft

1-supply duct; 2-fan; 3-exhaust duct; 4-fresh air; 5-air baffle; 6-exhaust fan; 7-jet fan; 8-polluted air

Firstly, the fresh air outside the tunnel is introduced into the positive tunnel through the inclined shaft by the exhaust fan, and the fresh air entering the positive tunnel is distributed according to the need by the jet fan. The polluted air flows from each heading face to the inlet of the exhaust fan through the tunnel, and then the exhaust fan is discharged to the outside of the tunnel through the exhaust duct.

2) Features

(1) Fresh air is transferred to the heading face.

(2) Most of the air flow in the inclined shaft and one tunnel is fresh.

(3) Soft air duct can be used in the supply duct, and the extension of the duct is

relatively easy.

(4) Leakage of exhaust duct has negative effect on ventilation and will cause secondary pollution.

(5) Large ventilation section and low power consumption.

(6) The requirement of air duct is small and the cost is low.

3) Key points for implementation

(1) Unused cross channels shall be closed in time. For construction purpose, air doors shall be installed for those that cannot be closed. If air doors cannot be installed, jet fans shall be used for regulation and control.

(2) Ventilation ducts shall be laid flat, straight and smooth, especially at corners without turning dead corners.

(3) The distance between the air outlet of the supply duct and the heading face is less than 5 times of the equivalent diameter of the positive tunnel.

(4) The cross channel of the exhaust fan shall be provided with the air separation door to reduce the usage of the jet fan. If the air separation door cannot be provided, the jet fan shall be increased for regulation.

(5) The air flow into the tunnel mainly depends on the air flow of the exhaust fan, and the exhaust fan selected must be able to meet the demand of the total air flow.

(6) The air outlet of the discharge duct shall be positioned above the tunnel opening at a distance of about 10 times the equivalent diameter of the inclined shaft or in a right-angle direction from the hole opening.

(7) The diameter of the exhaust duct shall be as large as possible, so that the exhaust fan with large air volume, low pressure head and relatively small power can be selected.

4. Parallel ventilation in the direction of entering the tunnel from the main and auxiliary inclined shafts

The two-way parallel construction of the main and auxiliary inclined shaft into the double-hole tunnel refers to the parallel construction of the main and auxiliary inclined shafts into the tunnel. The tunnel is constructed by two parallel double-holes, two directions and four working surfaces at the same time, and the two holes are connected by a cross channel with a certain distance, that is, the double-inclined shaft and double-hole tunnel method.

The transportation method is usually rail transportation, so the construction of rail transportation is considered here.

1) Ventilation method

When the blind heading distance is short, the system arrangement and key points of implementation are similar to those of the blind heading construction.

When the blind heading distance is long, jet gallery ventilation is adopted. The system arrangement is shown in Figure 1-29.

Chapter 1 Construction Ventilation of Underground Engineering Projects

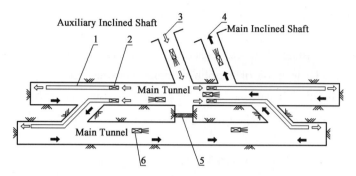

Figure 1-29 Schematic diagram of jet gallery ventilation system in a double-hole tunnel with double inclined shafts
1-supply duct;2-air blower;3-fresh air;4-polluted air;5-the damper;6-the jet machine

Fresh air is driven into the main tunnel from one inclined shaft, fresh air is sent to the heading face by four blowers and four ducts, polluted air flows from the heading face to the main inclined shaft through the tunnel, and finally discharged from the main inclined shaft.

2) Features

(1) Fresh air is transferred to the heading face.

(2) Most of the area of the auxiliary inclined shaft and a main tunnel is fresh air flow.

(3) Soft air duct can be used in the supply duct, and the extension of the duct is relatively easy.

(4) Large ventilation section and low cost.

(5) The requirement of air duct is small and the cost is low.

(6) Noise pollution is easily formed when the fan is arranged in the tunnel.

3) Key points for implementation

(1) The jet fan is preferably installed in a hole with a smaller cross-section.

(2) The unused cross channel shall be closed in time. For the construction needs, the air door shall be installed if it cannot be closed, and the jet fan shall be used to regulate and control the air door if it cannot be installed.

(3) Ventilation ducts shall be laid flat, straight and smooth, especially at corners without turning dead corners.

(4) The distance between the air outlet of the supply duct and the heading face is less than 5 times of the equivalent diameter of the positive tunnel.

(5) The fan must meet the noise standard requirements.

5. Ventilation method for two-way parallel construction from vertical shaft to tunnel

The two-way parallel construction from the vertical shaft to the tunnel is to set up a vertical shaft, enter the tunnel for construction, the tunnel is two parallel double holes, two directions, four working surface construction at the same time, between the two holes by a certain interval of cross channel connecting. People, materials and machines enter and exit from the vertical shaft, that is, the single vertical shaft double-hole tunnel method.

Rail transportation is usually used, so the ventilation for rail transportation is considered here.

1) Ventilation method

When the blind heading distance is short, the system arrangement and key points of implementation are similar to those of the blind heading construction.

When the blind heading distance is long, the positive pressure and exhaust air mixed type and jet tunnel type are adopted. The system arrangement is shown in Figure 1-30.

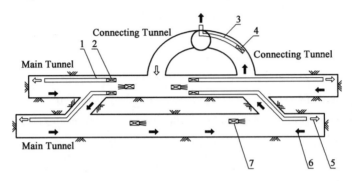

Figure 1-30 Schematic diagram of jet gallery ventilation system in double-hole tunnel with single shaft
1-supply duct; 2-air blower; 3-supply duct; 4-exhaust fan; 5-fresh air; 6-polluted air; 7-jet fan

Firstly, the fresh air outside the tunnel is introduced into the bottom of the vertical shaft through the vertical shaft and the connecting channel by the exhaust fan, then the fresh air is introduced into the main tunnel by the jet fan, and the fresh air is distributed and flowed according to the needs. Fresh air enters the main tunnel from the vertical shaft through the left connecting channel at the bottom of the vertical shaft, and is sent to the heading face through the supply duct by the blower, and polluted air converges from the heading face through the tunnel to the inlet of the exhaust fan arranged in the right connecting channel at the bottom of the vertical shaft, and then is discharged to the outside of the tunnel through the air exhaust duct by the exhaust fan.

2) Features

(1) Fresh air is transferred to the heading face.

(2) The vertical shaft is fresh air flow.

(3) Soft air duct can be used in the supply duct, and the extension of the duct is relatively easy.

(4) Large ventilation section and low power consumption.

(5) The requirement of air duct is small and the cost is low.

(6) Noise pollution is easily caused by the fan installed in the tunnel.

3) Key points for implementation

(1) The jet fan is preferably installed in a hole with a smaller cross-section.

(2) Unused cross channels shall be closed in time. For construction purpose, air doors shall be installed for those that cannot be closed. If air doors cannot be installed, jet fans shall be used for regulation and control.

(3) Ventilation ducts shall be laid flat, straight and smooth, especially at corners without

turning dead corners.

(4) The distance between the air outlet of the supply duct and the heading face is less than 5 times of the equivalent diameter of the tunnel.

(5) Set the air door in the right connecting channel of the exhaust fan to reduce the usage of the jet fan.

(6) The air volume entering the main tunnel is mainly determined by the air volume of the exhaust duct, and the selected exhaust fan capacity must be able to meet the demand of the total air volume.

(7) The diameter of the exhaust duct shall be as large as possible, so that the exhaust fan with large air volume, low pressure head and relatively small power can be selected.

(8) The fan must meet the noise standard requirements.

6. Ventilation method for two-way parallel construction from main and auxiliary vertical shafts into tunnel

Parallel construction from main and auxiliary shafts into double-hole means that two main and auxiliary shafts are set up to enter the main tunnel for construction. The main tunnel is two parallel double-hole, two directions and four working surfaces are constructed simultaneously, and the two tunnels are connected by a cross channel at a certain distance. Usually, the transportation method is rail transportation, that is, double shaft and double main tunnel method.

The ventilation method is approximately the same as that of the parallel tunneling of the double-vertical shaft and double-main tunnel.

7. Ventilation method of simultaneous construction of multiple working faces through inclined shaft

1) Ventilation method

Jet gallery ventilation is usually adopted, and the system arrangement is shown in Figure 1-31.

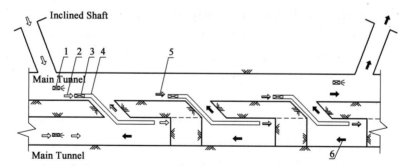

Figure 1-31 Schematic diagram of jet gallery ventilation for simultaneous construction of multiple working faces of double-hole tunnel with inclined shaft (track transportation)

1-jet fan;2-fresh air;3-air blower;4-supply duct;5-lightly polluted air;6-polluted air

A large enough main air flow is formed in a tunnel after the inclined shaft is penetrated by natural air and jet fans. The blowers are arranged separately on the windward side of each cross channel. The supply duct enters into different operation areas through each cross channel, and the air outlet is located near the heading face. Fresh air is sent to the heading face by the blower through the supply duct to dilute the contaminants, and polluted air is returned from the working surface to the cross channel, converged into the main air flow, and discharged out of the tunnel along the wind flow.

In special cases, such as trackless transportation, the centralized placement of blowers can be adopted, as shown in Figure 1-32.

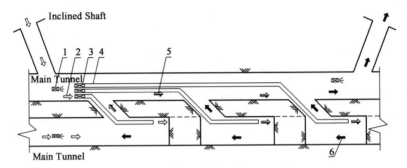

Figure 1-32 Schematic diagram of jet gallery ventilation for simultaneousconstruction of multiple working faces of double-hole tunnel with inclined shaft (track transportation)

1-jet fan;2-fresh air;3-air blower;4-supply duct;5-lightly polluted air;6-polluted air

2) Features

(1) Fresh air is transferred to the heading face.

(2) Soft air duct can be used in the supply duct, and the extension of the duct is relatively easy.

(3) Large ventilation section and low power consumption.

(4) Duct air leakage has positive effect on ventilation.

(5) When the blowers are placed separately, the blowers provided in the downdraft flow are sent to the working surface for air that is not completely fresh.

(6) Noise pollution is easily caused by the fan installed in the tunnel.

3) Key points for implementation

(1) The distribution of supply ducts shall be flat, straight and smooth, especially at corners without turning dead corners.

(2) The distance between the air outlet of the supply duct and the heading face is less than 5 times of the equivalent diameter of the positive tunnel.

(3) The jet fan shall be arranged in the inclined shaft with smaller cross section, and the ejecting direction shall be consistent with the natural wind direction.

(4) The main air flow shall be large enough to ensure the air intake quality of the blower.

(5) The blower must be arranged on the windward side of the cross channel.

(6) When the upstream blasting fume passes by, turn off the downstream blower temporarily.

(7) During trackless transportation, it is better for vehicles to enter and exit through the exhaust inclined shaft.

(8) The fan shall conform to the standard of noise specification.

1.5 Calculation of Construction Ventilation

1.5.1 Calculation of Air Volume

The calculation of air volume is mainly to calculate the amount of ventilation required under various conditions, such as personnel breathing, diluting the harmful gases emitted from surrounding rock, exhausting blasting fume, diluting diesel vehicle exhaust gas, exhausting dust and so on. Among these air volumes, if the requirements of maximum air volume can be met, the requirements of air volume for other projects can also be met in general. Therefore, the maximum air volume is selected as the required air volume of the tunnel.

1. Breathing air volume of operation personnel

The tunnel workers breathe out carbon dioxide (CO_2), which is also a kind of pollution to the tunnel working environment. When there are more operators, the pollution can not be ignored. The minimum ventilation required to breathe each operator can be calculated according to Formula (1-4):

$$Q_p = \frac{100c}{a - b} \tag{1-4}$$

Where: Q_p——Ventilation amount required for each operator to breathe (m^3/min);
a——Allowable concentration of CO_2 (%);
b——Allowable concentration of CO_2 in the atmosphere (%);
c——CO_2 exhaled by each operator (m^3/min).

Since $a = 0.5\%$, $b = 0.03\%$, $c = 1.2 \times 10^{-3} m^3/min$, so

$$Q_p = \frac{100 \times 1.2 \times 10^{-3}}{0.5 - 0.3} = 0.3 m^3/min \tag{1-5}$$

In this way, when ventilation is carried out only with the CO_2 exhaled by the operators, the air volume per operator is about $0.3 \ m^3/min$. The British code is based on this to determine the amount of air required by the operator. However, Japan believes that it is difficult to effectively ventilate the working environment with such a small air volume. Considering the environmental temperature and humidity, labor comfort, and the need for moderate air flow, the above air volume should be increased by 10 times. In our country, the tunnel construction usually adopts $3 \ m^3/min$ per person. Therefore, the total air volume required for the workers to breathe in the tunnel is as follows:

$$Q_p = 3N \tag{1-6}$$

Where: Q_p——Total ventilation volume required for workers to breathe in the tunnel (m^3/min);
N——Number of operators in the tunnel.

2. Air volume required for blasting fume exhaust

The blasting fume produced by blasting mainly includes toxic gases and dust such as carbon monoxide, carbon dioxide and nitrogen oxides. There are many formulas and methods to calculate the air volume needed to exhaust blasting fume after blasting, which is based on carbon monoxide.

1) Air volume required for fume extraction by supply ventilation blasting

(1) Voronin formula

When the distance from the outlet of the air duct to the working face is not greater than $(4 \sim 5) A^{0.5}$:

$$Q_b = \frac{0.456^3}{t} \sqrt{\frac{Gb(AL_0)^2}{P_q^2 C_a}} \tag{1-7}$$

Where: Q_b——Air volume required for blasting fume extraction (m^3/min);

t——Ventilation time (min);

G——Quantity of explosives blasted at the same time (kg);

b——CO generated per kilogram of explosives (L/kg), generally $b = 40$ L/kg, and $b = 100$ L/kg in case of coal seam;

A——Sectional area of tunnel excavation (m^2);

L_0——Ventilation length (m);

P_q——Ratio of the air volume at the beginning and end of the ventilation duct in the ventilation section;

C_a——Required CO concentration (%).

(2) Common formula

At present, the most commonly used formulas in the calculation of construction ventilation is:

$$Q_b = \frac{7.8^3}{t} \sqrt{G(AL_0)^2} \tag{1-8}$$

In fact, this formula is a simplified formula of Voronin formula when $b = 40$ L/kg, $C_a = 0.008\%$, $P_q = 1$. The biggest shortcoming of the formula is that the duct is not leaked. In fact, the pipe line near the working face is seriously damaged, the maintenance quality is low, and the air leakage is very large.

(3) Calculation method in "Calculation method of coal mine air volume"

The method is based on the fresh air volume of 500 m^3 needed to dilute the blasting fume after blasting per kilogram of explosives, that is:

$$Q_b = \frac{G \times 500}{t} \tag{1-9}$$

Where: t——Ventilation time (min);

G——Quantity of explosives blasted at the same time (kg).

(4) Recommended formula

For the above formulas, according to the field experience, the Voronin formula is

Chapter 1 Construction Ventilation of Underground Engineering Projects

recommended. However, C_a is the maximum allowable concentration (MAC) in the formula. When the required concentration standard is the short-time contact permission concentration (PC-STEL), that is, the weighted average concentration for 15 minutes, the formula becomes:

$$Q_b = \frac{0.456^3}{t}\sqrt{\frac{Gb(AL_0)^2}{P_q^2 C_{a15}} \frac{2t^2 + 15t}{2(t+15)^2}} \tag{1-10}$$

Where: C_{a15}——The 15 min weighted average concentration of CO required (%);

The other symbols are the same as before.

2) Exhaust air quantity required by exhaust ventilation blasting fume

For the exhaust ventilation with hard duct outside the tunnel, the calculation of air volume is mainly based on the Volonin method. When the distance from the end of the air duct to the working face is not more than $1.5 A^{0.5}$, the calculation formula is:

$$Q_b = \frac{0.254}{t}\sqrt{\frac{GbAL_t}{C_a}} \tag{1-11}$$

Where: L_t——Fume throwing length (m).

When $b = 40$ L/kg and $C_a = 0.008\%$, the formula becomes:

$$Q_b = \frac{18}{t}\sqrt{GSL_t} \tag{1-12}$$

Where the symbolic meaning is the same as before.

Similarly, where C_a is the maximum allowable concentration (MAC) when the required concentration standard is the short-time contact permission concentration (PC-STEL), that is, the weighted average concentration for 15 minutes, the formula is:

$$Q_b = \frac{0.254}{t}\sqrt{\frac{GbAL_t}{C_a} \frac{t}{t+15}} \tag{1-13}$$

3) Air volume required for fume extraction by compound ventilation blasting

(1) Voronin formula

The fume exhaust process of the compound ventilation is basically the same as that of the supply ventilation. The difference is that the blasting fume enters the exhaust duct when it is exhausted to the inlet of the exhaust duct. The formula for calculating the fume exhaust air volume is as follows:

$$Q_b = \frac{0.456}{t}\sqrt{\frac{Gb(AL_0)^2}{P_q^2 C_a}} \tag{1-14}$$

Where the symbolic meaning is the same as before.

(2) Japanese calculation formula

$$Q_b = 0.368 \frac{P}{Rat} \tag{1-15}$$

Where: P——Amount of harmful substances generated after blasting (m³);

R——Ventilation efficiency.

3. Calculate the air volume at the allowable minimum wind speed

Dust is mostly produced by concrete spraying and slag loading and transportation. Fume emission from vehicles is also a factor. If the amount of dust occurs is known, it is very simple to calculate the air volume of the dust, which can be calculated according to the following formula:

$$Q_d = \frac{M}{n} \tag{1-16}$$

Where: Q_d——Air volume required for dust removal from the working surface (m^3/min);

M——Amount of respirable dust generated on the working surface (mg/min);

n——Allowable dust content (mg/min).

Because it is difficult to determine the amount of dust produced and to calculate the ventilation rate with dust as the object, it can be calculated on the basis of the correlation between the wind speed in the tunnel and the dust concentration in the tunnel outside the working face. For example, the wind speed in the tunnel is about 0.3 m/s and the dust concentration can be diluted to less than 2 mg/m^3. Therefore, the wind speed in foreign countries is more than 0.3 m/s.

4. Calculation of supply based on dilution and exhaust of exhaust gases from internal combustion engines

When using internal combustion engine power equipment, the ventilation amount of the tunnel shall be sufficient to dilute and discharge all the exhaust gas discharged from the equipment, so that the concentration of harmful gases in the air of the main working sites in the tunnel shall be reduced below the allowable level.

1) Calculated according to tunnel construction code

According to the tunnel construction code, the total air volume required to dilute the exhaust gas of internal combustion equipment is:

$$Q_s = 3 \sum_{i=1}^{n} N_i \tag{1-17}$$

Where: Q_s——Total air volume required to dilute the exhaust gas of the internal combustion equipment (m^3/min);

N_i——Rated power of each internal combustion device (kW).

2) Relevant provisions of the United Kingdom

The British BS6164—2001 suggest that the supply should be 9 m^3/(min · kW) according to the cross-sectional area of the tunnel, plus 1.9 m^3/(min · kW) for diesel engines. It is recommended to provide at least 3.0 m^3/(min · kW) of fresh air for machinery with strict emission control over a period of time when supplying air to mechanical equipment, especially diesel engine equipment.

3) Relevant provisions of South Africa

The minimum supply volume is 6.0 m^3/(min · kW) when the rated power of the internal combustion engine is used.

4) Relevant provisions of the International Tunneling Association

The International Tunneling Association (ITA) requires a minimum supply of 4 m³/(min · kW) for diesel machinery based on rated power.

5) Recommendation

According to the actual situation of vehicles in our country, it is suggested that the supply standard should be 4.5 m³/min per kilowatt rated power, and the minimum supply should be 4 m³/min per kilowatt according to the standard of International Tunneling Association (ITA), that is, the minimum supply should be 4 m³/min per kilowatt.

5. Calculate the air volume according to the gas emission

If there is gas emission from the working face, sufficient air must be supplied to the working face to dilute and discharge the gas, so as to ensure that the gas concentration is below the allowable concentration. That is:

$$Q_g = \frac{100 q_{CH_4}}{C_a - C_0} K \qquad (1\text{-}18)$$

Where: Q_g——Air volume required to discharge gas (m³/min);

q_{CH_4}——Aas emission from working face (m³/min);

C_a——Allowable gas concentration in the working face, 1%;

C_0——Concentration of gas in the airflow sent to the working face;

K——Unbalanced coefficient of gas emission, $K = 1.5 \sim 2$.

1.5.2 Calculation of Duct Air Leakage

In the blind heading construction, duct ventilation is often used to deliver fresh air to the construction work surface, exhaust harmful gas and dust from working face to create necessary working environment to meet the needs of construction. Duct air leakage is the main problem of duct ventilation. Duct air leakage rate is the main standard to evaluate the quality of duct installation and is one of the main basis to determine the supply of fan.

1. Approximate calculation by applying the theory of average air leakage rate of hectometer

The average hectometer air leakage rate refers to the percentage of the average air leakage per hectometer duct to the fan supply Q_f.

$$P_{100} = \frac{Q_f - Q_0}{Q_f - L\%} \times 100\% \qquad (1\text{-}19)$$

Where: P_{100}——The average hectometer air leakage rate of the duct;

Q_f——Fan supply (m³/min);

Q_0——Air volume at the end of the duct (m³/min);

L——Duct length (m).

The essence of the formula is to assume that the air leakage is the same in every hectometer. Total air leakage of air duct:

$$q_L = Q_f P_{100} \frac{L}{100} \tag{1-20}$$

2. Approximate calculation by using Takagi Hideo theory

$$Q_0 = Q_f e^{-ZL} \tag{1-21}$$

Where: L——Duct length (m);

The other symbols are the same as before.

The essence of the formula is that the air leakage rate of hectometer in every part of the duct is a fixed value.

3. Approximate calculation by using the theory of Qinghan Tunnel

$$Q_f = \frac{Q_0}{(1-\beta)^{\frac{L}{100}}} \tag{1-22}$$

Where: β——the average air leakage rate of hectometer;

The other symbols are the same as before.

The essence of the formula is that the air leakage rate of hectometer in every part of the duct is a fixed value.

4. Calculate the air leakage rate of hectometer at the end of the duct

The air leakage rate at the end of the duct depends only on the ratio of the air leakage resistance to the ventilation resistance at the end of the duct, and is independent of the air volume in the duct. The formula for calculating the hectometer air leakage rate at the end of the duct is as follows:

$$Q_f = \left[1 + \frac{1-\sqrt{1-M_{100}}}{\sqrt{1-M_{100}}}\left(\frac{L}{100}\right)^{\frac{3}{2}}\right]^2 \tag{1-23}$$

Where: M_{100}——Air leakage rate at the end of hectometer;

The other symbols are the same as before.

5. Approximate calculation using Voronin's theory

For hard duct:

$$Q_f = \phi Q_0 \tag{1-24}$$

$$\phi = 1 + \frac{1}{3} dmk \sqrt{R_0} L^{\frac{3}{2}} \tag{1-25}$$

Where: ϕ——Reserve factor of air leakage of air duct;

d——Duct diameter (m);

R_0——Diameter of air duct (m);

m——Number of pipe fittings per unit length of duct (m);

k——Air leakage coefficient of each joint of the air duct with a diameter of 1 m;

The other symbols are the same as before.

1.5.3 Calculation of Duct Ventilation Resistance

Ventilation resistance is the main basis for selecting fans, including friction resistance and local resistance.

1. Frictional resistance

When there is no air leakage in the duct, the calculation formula is:

$$h_f = \lambda \frac{L}{d} \frac{v^2}{2} \rho \quad (1\text{-}26)$$

Where: h_f——Frictional resistance of duct (Pa);
λ——Coefficient of friction;
L——Duct length (m);
d——Duct diameter (m);
v——Velocity of air flow in the duct (m/s);
ρ——Air density (kg/m^3).

When there is no air leakage in the duct, the calculation formula is:

$$h_f = \frac{400 \lambda \rho}{\pi^2 d^5} \frac{1 - (1-\beta)^{-\frac{2L}{100}}}{\ln(1-\beta)} Q_0^2 \quad (1\text{-}27)$$

Where: h_f——Frictional resistance of duct (Pa);
λ——Coefficient of friction;
L——Length of air duct (m);
d——Equivalent diameter of the cross-section through the wind (m);
β——Air leakage rate of hectometer of air duct;
Q_0——Air volume at fan working point (m^3/s);
The other symbols are the same as before.

2. Local resistance

The energy consumption will be generated when the air flow passes through the ducts with sudden expansion or contraction, turning and crossing, etc. The calculation formula is as follows.

1) Local resistance to sudden expansion or contraction

$$h_x = \xi \frac{v^2}{2} \rho \quad (1\text{-}28)$$

Where: h_x——Local resistance of the duct (Pa);
ξ——Coefficient of friction;
v——Wind speed at the small section of the duct (m/s);
ρ——Air density (kg/m^3).

2) Other local resistance

The other local resistance can be calculated by the relevant table to find out the local resistance coefficient, and then calculate the corresponding wind speed and cross-sectional area can be.

3) Ventilation resistance of tunnel

Because the cross-sectional area of the tunnel is much larger than the cross-sectional area of the ventilation duct, the ventilation resistance of the tunnel cannot be ignored when the diameter of the ventilation duct is larger than the cross-sectional area of the tunnel, or when the tunnel is very long.

1.6 Construction Ventilation Equipment and Selection

Construction ventilation equipment is mainly the fan and ventilation pipe, the fan is the power source of the ventilation system, ventilation duct is the ventilation system of the air flow channel.

1.6.1 Ventilator

1. Types of ventilators

Fan is our country to the gas compression and the gas transportation machinery custom abbreviation, it includes the ventilator, the blower and the compressor, their difference lies in the outlet air pressure is different. The air pressure of the ventilator shall not be less than 15 kPa. The air pressure of the blower is 15-340 kPa; the air pressure of the compressor is greater than 340 kPa.

Ventilator is a kind of fan, but people usually call it fan for short. According to the direction of gas flow, the fan can be divided into centrifugal, axial, oblique and cross flow types.

When the centrifugal fan works, the impeller is driven by the power machine to run in the volute casing, and air is sucked in from the center of the impeller through the suction port. Due to the dynamic action of the blade on the gas, the gas pressure and speed are increased, and the gas is thrown along the blade passage to the casing under the action of centrifugal force, and discharged from the exhaust port. According to the pressure, the centrifugal fan can be divided into: low pressure centrifugal fan (less than 1 kPa); medium pressure centrifugal fan (1-3 kPa); high pressure centrifugal fan (3-15 kPa).

When the axial-flow fan works, the impeller is driven by the power machine to rotate in the cylindrical casing, gas enters from the current collector, energy is obtained through the impeller, pressure and speed are increased, and then the impeller is discharged along the axial direction. Axial flow fans can be divided into: low pressure axial flow fans (less than 0.5 kPa); high pressure axial fan (0.5-15 kPa).

Oblique flow fan is also called mixed flow fan. In this kind of fan, the gas enters the impeller at an angle with the axis, obtains energy in the impeller passage, and flows out in the oblique direction. This kind of fan has the characteristics of both centrifugal and axial flow, the flow range and efficiency are between the two.

Cross-flow fan is a small high-pressure centrifugal fan with forward multi-wing impeller. The gas enters the impeller from one side of the outer edge of the rotor and then exits the impeller from the other side through the interior of the impeller, where it is subjected to twice the force of the blades. Under the condition of the same performance, its size is small and its rotation speed is low.

2. Types of ventilators for tunnel construction

Most of the ventilators used in tunnel construction are axial flow fans. Because of its low air pressure, large air volume and convenient connection, it is widely used in tunnel construction ventilation. Centrifugal fans are rarely used in tunnel construction ventilation.

1) Categories by rotation speed

(1) **Single-speed fan**. Fans with fixed rotational speed are called single-speed fans.

(2) **Variable speed fan**. Variable speed fans are called variable speed fans. According to the principle of variable speed can be divided into variable pole speed fan and variable frequency speed fan. The variable-pole adjustable-speed fan adjusts the speed by changing the number of pole pairs of the motor, which is a step-by-step adjustable-speed fan. Variable frequency variable speed fan is to change the frequency of the motor power supply to achieve speed regulation, which is stepless speed regulation.

2) Categories by series

(1) **Single-stage fan**. A fan consisting of a set of blades and a motor is called a single-stage fan.

(2) **Two-stage fan**. Fans composed of two single-stage fans in series are called two-stage fans. According to the arrangement of the two-stage impellers and the direction of rotation, the two-stage impellers are divided into two types: the counter-cyclone and the non-counter-cyclone.

(3) **Multi-stage fan**. Fans with more than three single-stage fans in series are called multi-stage fans. If the single-stage fan is twofold, it is called counter-rotating multi-stage fan. Otherwise, it is non-counter-rotating multi-stage fan.

3) Categories by purpose

(1) **Main ventilator**. In tunnel ventilation, the main fan is used for the whole tunnel, which provides all the air volume to the tunnel through the tunnel. Mine ventilation is called the main fan, referred to as the main fan.

(2) **Local ventilator**. In roadway ventilation, the fan used to supply air to the working face through the ventilation duct is called the local fan. The mine ventilation is called local fan, abbreviated as local fan.

(3) **Jet fan**. In the tunnel ventilation, the high-speed jet fan is used to induce the directional flow of air in the tunnel or channel, which is called jet fan.

3. Fan performance

1) Fan performance parameters

The main parameters of fan performance are air volume Q, air pressure H, fan shaft power N and efficiency, etc.

(1) Fan flow

The flow rate of the fan generally refers to the volume of air passing through the inlet of the fan per unit time, also known as the volume flow rate (in the standard state unless otherwise specified), in units of m³/h, m³/min or m³/s.

(2) Fan (actual) full pressure H_t and static pressure H_s

The full pressure H_t of the fan is the energy that the fan works on the air and consumes 1 cubic meter of air in units of N · m/m³ or Pa, the value of which is the difference between the full pressure of the outlet air flow of the fan and the full pressure of the inlet air flow. When the natural air pressure is neglected, H_t is used to overcome the resistance h_R of the ventilation network and the kinetic energy loss h_v of the fan outlet for the negative pressure exhaust ventilation. The air pressure overcoming the resistance of the pipe network is called the static pressure H_s of the fan, that is, the total pressure of the fan is equal to the sum of the static pressure and the kinetic energy loss at the outlet of the fan.

(3) Power of the fan

When the output power of the fan (also called air power) is calculated at full pressure, it is called full pressure power N_t(kW), which is calculated by the following equation:

$$N_t = \frac{H_t Q}{1000} \quad (1\text{-}29)$$

Calculating the output power with a static pressure fan is referred to as static pressure N_s, that is:

$$N_s = \frac{H_s Q}{1000} \quad (1\text{-}30)$$

Therefore, the shaft power of the fan, that is, the input power N (kW) of the fan:

$$N = \frac{N_t}{\eta_t} = \frac{H_t Q}{1000 \eta_t} \quad (1\text{-}31)$$

Or

$$N = \frac{N_s}{\eta_s} = \frac{H_s Q}{1000 \eta_s} \quad (1\text{-}32)$$

Where: η_t, η_s——Fan full pressure and static pressure efficiency.

(4) Fan efficiency

Fan efficiency is the percentage of the ratio of the output power to the input power of the fan. Reflect the degree of effective utilization of shaft power in the process of energy transmission of the fan, expressed in terms of η_s. The percentage of the ratio of full voltage power N_t to input power N is called full voltage efficiency. The percentage of the ratio of the static pressure power N_s to the input power N is referred to as the static pressure efficiency.

$$\eta_t = \frac{N_t}{N} \times 100\% \quad (1\text{-}33)$$

$$\eta_s = \frac{N_s}{N} \times 100\% \quad (1\text{-}34)$$

2) Individual characteristic curve of the fan

When the fan works at a certain speed in the duct of wind resistance R, a set of working parameters, such as air pressure H, air volume Q, power N and efficiency η, can be calculated, which is the operating point of the fan when the wind resistance of the duct is R. By changing the wind resistance of the duct, another set of corresponding working parameters can be obtained. By changing the wind resistance of the duct many times, a series of operating parameters can be obtained. These parameters are correspondingly depicted in a rectangular coordinate system with Q as the abscissa and H, N and η as the ordinate, and the points with the same name are connected by smooth curves, that is, H-Q, N-Q and η-Q curves, which are called individual characteristic curves of the ventilator under the condition of the rotational speed. .

Figure 1-33 and Figure 1-34 are examples of individual characteristic curves of axial and centrifugal fans, respectively. Generally, there are saddle humps in the air pressure characteristic curve of axial fan. And the hump area of the same ventilator increases with the increase of blade angle. The hump point D is a stable working area with the right characteristic curve as a monotonic descending section. Point D is an unstable working area to the left. When the fan is working in this area, the air volume, air pressure and motor power of the fan will fluctuate sharply, even the airframe will vibrate and make abnormal noise, which will destroy the fan if it is serious. The hump of the air pressure curve of the centrifugal fan is not obvious and gradually decreases with the increase of the blade caster angle, and the working area of the air pressure curve of the centrifugal fan is flatter than that of the axial fan. When the change of the wind resistance is the same, the air amount of the centrifugal fan changes greater than that of the axial-flow fan.

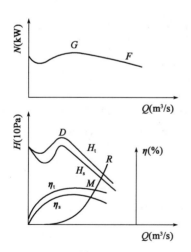
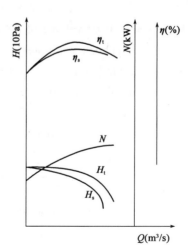

Figure 1-33 Individual characteristic curve of axial flow fan   Figure 1-34 Individual characteristic curve of centrifugal fan

3) Reasonable range of fan operating points

In order to make the ventilator run safely and economically, it is necessary to put the operating points within a reasonable range.

From the point of economic view, the operation efficiency of the fan should not be lower than 60%. In terms of safety, the operating points of the fan must be located on the lower right side and the monotonously descending section of the hump. Because the performance curve of axial flow fan has saddle-shaped section, in order to prevent the occasional increase of wind resistance and other reasons, it is necessary to limit the actual working wind pressure not to exceed 90% of the maximum wind pressure.

4) Series and parallel of fans

When the air volume of the fan needs to be increased, the method of paralleling several wind turbines can be adopted; when static pressure needs to be increased, several wind turbines can be connected in series. These methods should carefully study the resistance of the air duct and the characteristics of the fan in order to prevent the failure of the intended purpose.

(1) Series operation of fans

The air inlet of one typhoon machine is connected to the air outlet of another typhoon machine directly or through a section of pipe (or tunnel) at the same time, which is called series operation of fans. There are centralized series, spaced series and disconnected series.

The characteristic of fan working in series is that the total air flow through the duct is equal to the air flow of each wind turbine (no air leakage), and the sum of the working air pressure of the two wind turbines is equal to the resistance of the duct. That is:

$$h = H_{s1} + H_{s2} \qquad (1\text{-}35)$$

$$Q = Q_1 = Q_2 \qquad (1\text{-}36)$$

Where: h——The resistance of the pipe network (Pa);

Q——The total air volume of the pipe network (m³/s);

H_{s1}, H_{s2}——1, 2 typhoon static pressure (Pa);

Q_1, Q_2——1, 2 wind volume of the two wind turbines (Pa).

While disconnecting the series does not.

Centralized series as shown in Figure 1-35, the manner in which the fans F_1 and F_2 are directly connected in series without any ventilation ducts is referred to as centralized series.

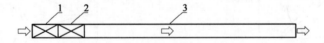

Figure 1-35 Centralized series schematics of fans

1-fan F_1; 2-fan F_2; 3-duct L

As shown in Figure 1-36, the method of operation in which the blowers F_1 and F_2 are connected in series through a section of pipe L_0 is referred to as indirect series connection.

Disconnect series as shown in Figure 1-37, the fans F_1 and F_2 are disconnected in series in a target operating method through a section of piping, called disconnected series.

(2) Parallel operation of fans

As shown in Figures 1-38 and 1-39, the operation method in which the air inlets of the two wind turbines are connected directly through a section of a tunnel is called parallel ventilation. Parallel fans are divided into centralized parallel and diagonal parallel. Figure 1-38 shows the centralized parallel connection, and Figure 1-39 shows the diagonal parallel connection.

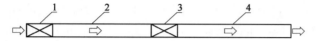

Figure 1-36 Interval series schematics of fans
1-fan F_1 ;2-duct Lo;3-fan F_2 ;4-duct L

Figure 1-37 Disconnected series schematics of fans
1-fan F_1 ;2-duct L_0 ;3-fan F_2 ;4-duct L

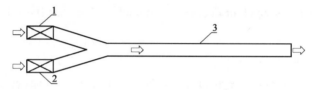

Figure 1-38 Centralized parallel schematics
1-fan F_1 ;2-fan F_2 ;3-duct L

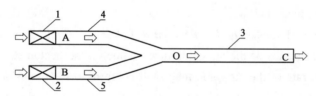

Figure 1-39 Diagonal parallel schematics
1-fan F_1 ;2-fan F_2 ;3-duct OC;4-duct AO;5- duct BO

Centralized parallel operation is characterized in that the air inlets (or outlets) of the two wind turbines are considered to be connected at the same point, the static pressure of the two wind turbines is equal to the resistance of the duct, and the air flow through the duct is equal to the sum of the air flow of the two wind turbines. That is:

$$h = H_{s1} = H_{s2} \tag{1-37}$$
$$Q = Q_1 + Q_2 \tag{1-38}$$

The symbolic meaning in the formula is the same as before.

In diagonal parallel connection, the resistance of the line OA and the line OB is subtracted from the fan F_1 and the fan F_2, respectively, so that the fan F_4 and the fan F_5 can be equivalent to each other, so that the fan F_4 and the fan F_5 are parallel to the point O, and the working characteristics are the same as those of the centralized parallel connection.

1.6.2 Air Duct

1. The type of air duct

The air ducts commonly used in tunnel construction ventilation are soft air ducts, hard air ducts and flexible air ducts.

Soft air duct is made of coated or dipped plastic cloth. The base cloth is polyester, vinylon and other textiles, which has small quality and is easy disassembly and handling with small air leakage, suitable for large diameter air duct. But soft air duct can only be used for positive pressure ventilation.

Hard air ducts are generally iron air ducts and are usually connected by flanges. There are also FRP air duct, aluminum plastic duct and glass fiber plastic duct overseas. They usually are the plates processed and molded on site. It can be used for positive or negative pressure ventilation.

Flexible air duct is usually made of soft air duct and steel ring. It is mainly used for negative pressure ventilation and short-distance ventilation because of high resistance.

In addition, there is a special air duct for gas tunnels, which has flame retardant and antistatic properties.

2. Performance of air ducts

The performance indexes of the air duct are mainly air leakage rate, wind resistance, friction coefficient (Darcy coefficient), pressure resistance and diameter change rate. At present, zipper soft air duct is the most widely used in tunnel construction ventilation, so the performance of soft air duct is mainly introduced below.

The national standard *Rubber or Plastic Coated Fabric Air Guide Tube* (GB/T 9900-2008) gives the performance indexes of the air guide tube, namely, the hectometer wind resistance and hectometer air leakage rate of the air guide tube shall conform to the provisions of Table 1-9, and the pressure resistance and diameter change rate of the air guide tube shall conform to the provisions of Table 1-10. The hectometer wind resistance and hectometer air leakage rate were measured according to the Method for Determination of *Air Leakage Rate and Wind resistance of Air Tubes* (GB/T 15335—2006), while the pressure resistance and diameter change rate were determined according to *Method for Determination of Tensile Length and Break Elongation of Rubber or Plastic Coated Fabrics* (HG/T 2580—2008).

Hectometer wind resistance and Hectometer air leakage rate of air guide tube   Table 1-9

Air guide tube diameter (mm)	Positive pressure air guide tube		Negative pressure air guide tube	
	Hectometer wind resistance ($N \cdot S^2/m^8$)	Hectometer air leakage ratio (%)	Hectometer wind resistance ($N \cdot S^2/m^8$)	Hectometer air leakage ratio (%)
300	≤811	≤4	≤1728	≤5
400	≤196		≤410	

continued

Air guide tube diameter (mm)	Positive pressure air guide tube		Negative pressure air guide tube	
	Hectometer wind resistance ($N \cdot S^2/m^8$)	Hectometer air leakage ratio (%)	Hectometer wind resistance ($N \cdot S^2/m^8$)	Hectometer air leakage ratio (%)
500	≤54	≤4	≤134	≤5
600	≤24		≤54	
800	≤6		≤13	
1000	≤2		≤4	
1200	≤1		≤1.5	
1200 >	≤1		≤1	

Pressure resistance and diameter change rate of air guide tube   Table 1-10

Air guide tube diameter (mm)	Positive pressure air guide tube		Negative pressure air guide tube	
	Wind pressure (positive) (Pa)	Hectometer expansion ratio (%)	Wind pressure (negative) (Pa)	Hectometer expansion ratio (%)
300-500	>5100	≤3	>5000	≤3
600 and above	>5100	≤3	>4000	≤3

Hectometer wind resistance:

$$R_{100} = \frac{100\alpha U}{A^3} = \frac{6400\alpha}{\pi^2 d^5} \quad (1-39)$$

$$\alpha = \frac{\rho}{8}\lambda \quad (1-40)$$

Where: R_{100}——Hectometer wind resistance ($N \cdot s^2/m^8$);
d——The diameter of the ventilation duct (m);
λ——friction coefficient;
A——The cross-sectional area (m^2);
ρ——Air density (kg/m^3);
α——friction resistance coefficient ($N \cdot s^2/m^4$);
U——hydraulic perimeter (m).

It can be seen that the hectometer wind resistance is inversely proportional to the fifth power of the diameter of the air duct and directly proportional to the coefficient of friction resistance, which reflects the resistance characteristics of the air duct. According to the above formulas and Table 1-9, the friction resistance coefficient and friction coefficient standards corresponding to different diameters and hectometer wind resistance can be calculated, as shown in Table 1-11.

The friction coefficient (Darcy coefficient) λ is mainly determined by the relative smoothness of the inner wall of the air duct, which has a great influence on the ventilation resistance. Switzerland divides the air ducts into three grades: S, A and B. The grade λ is different for different grades, as shown in Table 1-12.

Relationship between hectometer wind resistance and friction resistance   Table 1-11
coefficients of air guide tubes

Air guide tube diameter (mm)	Positive pressure air guide tube			Negative pressure air guide tube		
	Hectometer wind resistance ($N \cdot s^2/m^8$)	Friction factor ($N \cdot s^2/m^4$)	Friction coefficient	Hectometer wind resistance ($N \cdot s^2/m^8$)	Friction factor ($N \cdot s^2/m^4$)	Friction coefficient
300	811	0.0030	0.0203	1 728	0.0065	0.0432
400	196	0.0031	0.0206	410	0.0065	0.0432
500	54	0.0026	0.0173	134	0.0065	0.0430
600	24	0.0029	0.0192	54	0.0065	0.0432
8000	6	0.0030	0.0202	13	0.0066	0.0434
1 000	2	0.0031	0.0206	4	0.0062	0.0411
1200	1	0.0038	0.0256	1.5	0.0058	0.0384

Friction coefficient   Table 1-12

Duct grade	Friction factor	Effective air leakage area f^* (mm^2/m^2)
S	0.015	5
A	0.018	10
B	0.024	20

1.6.3 Selection of Ventilation Equipment

Air ducts are usually selected firstly for ventilation. Then, the ventilation fans are considered. Sometimes it is necessary to make appropriate adjustment to the selected air ducts according to the selected fans.

1. Selection of air duct

Selection principles of air duct:

(1) The diameter of the air duct shall be such that the air volume of the fan can meet the air demand of the working surface at the maximum supply distance.

(2) Under the condition of tunnel section permission, select the air duct with larger diameter as far as possible to reduce ventilation resistance and save ventilation energy consumption.

(3) The air leakage rate and friction coefficient of hectometer air duct should be small.

(4) Easy to handle, install and maintain, strong and durable.

(5) The gas tunnel shall also have flame retardant and antistatic properties.

2. Selection of fans

1) Main Principles for Selection of Fans

(1) The fan shall be able to meet the supply requirements of the maximum supply distance.

(2) The working point of the fan shall not be in the surge area, but shall be within a

reasonable working range and as close as possible to the highest efficiency point.

(3) Choose low-noise, high-efficiency and energy-saving fan.

(4) Explosion-proof fan should be selected for gas tunnel.

2) Step of selecting the fans

The selection of the fans is closely related to the selection of the air duct. After the selection of the air duct, the selection of the fan should follow the steps:

(1) Calculate the required fan outlet air volume and air pressure according to the required ventilation volume, the air leakage and friction coefficient of the air duct, and the maximum supply distance.

$$Q = F_Q(x_1, x_2, x_3, x_4) \tag{1-41}$$

$$H_t = F_H(x_1, x_2, x_3, x_4) \tag{1-42}$$

(2) Calculate the effective power of the required fan according to the supply volume and air pressure of the fan.

$$N_t = QH_t \tag{1-43}$$

Where: N_t——effective power of the fan (W);

Q——air volume (m³/s);

H_t——wind pressure (Pa).

(3) According to the full pressure efficiency of the fan, the efficiency of the motor and the transmission efficiency, the input power of the motor is calculated.

$$N_m = \frac{QH_t}{\eta_t \eta_m \eta_{tr}} \tag{1-44}$$

Where: N_m——effective power of the fan (W);

Q——air volume (m³/s);

η_t——full pressure efficiency of the fan;

η_m——the efficiency of the motor;

η_{tr}——the transmission efficiency;

H_t——wind pressure (Pa).

(4) Determine the alternative fans according to the required fan supply volume, air pressure and power.

(5) Determine the working point according to the characteristic curve of the alternative fan and the characteristic curve of the air duct.

(6) Calculate the effective air volume on the working surface according to the air volume at the working point of the fan and the air pressure, and see if the air volume can meet the requirements. If the air volume meets the requirements, determine the selected fan.

Exercise

1.1 What are the common harmful gases in the air during construction of underground space?

1.2 What are the provisions of relevant codes for underground engineering in various industries regarding the working environment in the process of underground engineering construction?

1.3 What are the necessary factors for the formation of natural air flow in the tunnel?

1.4 What are the influencing factors of natural wind pressure?

1.5 What are the common natural ventilation methods during tunnel construction? What are their characteristics?

1.6 What are the basic mechanical ventilation methods during tunnel construction? What are their characteristics?

 Key vocabulary:

axial-flow fan　轴流风机
blowing ventilation　压入式通风
compound ventilation　混合式通风
dynamic pressure　动压
frictional resistance　摩擦阻力
hectometer wind resistance　百米风阻
inclined shaft　斜井
local resistance　局部阻力
natural ventilation　自然通风
pressure gradient　气压梯度
single-tube tunnel　单洞隧道
thermal potential difference　热位差
twin-tube tunnel　双洞隧道
vertical shaft　竖井

blind heading　独头掘进
centrifugal fan　离心风机
diagonal fan　斜流风机
exhaust ventilation　排出式通风
gallery ventilation　巷道式通风
horizontal pressure difference　水平压差
jet fan　射流风机
mechanical ventilation　机械通风
parallel adit　平行导坑
service gallery　辅助坑道
static pressure　静压
trackless transportation　无轨运输
ventilation resistance　通风阻力

Chapter 2 Operation Ventilation of Highway Tunnels

[**Important and Difficult Contents of this Chapter**]

(1) Harm of noxious gases in highway tunnel and hygienic standard of highway tunnel operation.

(2) Calculate the air demand of highway tunnel from three aspects: diluting fume, diluting carbon monoxide and ventilating demand.

(3) Various operation ventilation methods and their characteristics of highway tunnels.

(4) Pressure pattern and calculation method of longitudinal ventilation.

(5) The principle of selecting the ventilation method of highway tunnel.

2.1 Hygienic Standard for Operation of Highway Tunnels

During the driving process of the vehicle, the concentration of harmful gases in the tunnel is increasing, and when the concentration of CO and NO_2 reaches a certain value, the driver and occupant will be unwell. Too high a concentration of fume will reduce visibility in the tunnel and affect traffic safety. Therefore, it is necessary to carry out tunnel ventilation so that the air composition can meet the requirements of safety, health and comfort in the tunnel. In China, *Guidelines for Design of Ventilation of Highway Tunnels* (*JTG/T D70/2-02—2014*) stipulates that the design safety standard of highway tunnels is mainly to dilute the smoke and dust emitted by motor vehicles; Dilution of carbon monoxide (CO) emitted from motor vehicles is the main driving hygiene standard for highway tunnels, and dilution of nitrogen dioxide (NO_2) may be considered if necessary. The comfort standard of highway tunnel design is mainly the peculiar smell caused by ventilation and dilution of motor vehicles.

The toxicity of common harmful gases in highway tunnels is described as follows:

(1) Carbon monoxide has a strong affinity for hemoglobin, which binds to hemoglobin to form carboxyhemoglobin, resulting in the loss of oxygen-carrying capacity of hemoglobin, resulting in hypoxia in various tissues of the body.

(2) Nitrogen oxides are dissolved into nitrite and nitric acid in the deep part of respiratory system after inhalation, which is irritating and can cause congestion of throat and bronchus,

neutralize nitrate and nitrite with alkalis in cell tissue, cause arterial dilatation, decrease blood pressure, and cause headache and dizziness.

(3) Sulfurous acid is formed when sulfur dioxide is inhaled and dissolved by human body, which has strong irritation to upper respiratory tract and eyes.

(4) Fume contains incomplete burning hydrocarbons, which can irritate the throat and respiratory tract after inhalation. Another hazard of soot is that it affects visibility and driver vision in tunnels.

(5) Aldehydes, including formaldehyde and acetaldehyde, irritate eyes and respiratory system, and have bad odor.

2.1.1 Design concentration of fume

The value of design concentration K of fume in the tunnel shall conform to the following provisions:

(1) When the sodium light source with color rendering index $33 \leqslant R_a \leqslant 60$ and relevant color temperature 2000-3000 K is used, the fume design concentration K shall be taken as per Table 2-1.

Design concentration of fume K (Sodium Light Source)   Table 2-1

Design speed v_t/(km/h)	$\geqslant 90$	$60 \leqslant v_t \leqslant 90$	$50 \leqslant v_t < 90$	$30 < v_t < 50$	$v_t \leqslant 30$
Fume design concentration k/m^{-1}	0.0065	0.0070	0.0075	0.0090	0.0120*

Note: ① Measures such as traffic control or tunnel closure should be taken under this condition.
② When using fluorescent lamp, LED lamp and other light sources with color development index $R_a \geqslant 65$ and relevant color temperature of 3,300-6,000 K, the fume design concentration K shall be taken as per Table 2-2.

(2) When using a fluorescent lamp such as a fluorescent lamp or an LED lamp with a color rendering index $R_a \geqslant 65$ and a correlated color temperature of 3300-6000 K, the fume design concentration K shall be in accordance with Table 2-2.

Design Concentration of fume k (Light sources such as fluorescent lamps and LED lamps)   Table 2-2

Design speed v_t(km/h)	$\geqslant 90$	$60 \leqslant v_t \leqslant 90$	$50 \leqslant v_t < 90$	$30 < v_t < 50$	$v_t \leqslant 30$
Fume design concentration k/m^{-1}	0.0050	0.0065	0.0070	0.0075	0.0120*

Note: ① Traffic control or tunnel closure measures should be taken under this condition.
② When one-way traffic in two tunnels is temporarily changed to two-way traffic in one tunnel, the allowable concentration of fume in the tunnel should not be greater than 0.012, 4 to m^{-1}.
③ During the maintenance of the tunnel, the allowable fume concentration of the air support in the working section of the tunnel shall not be greater than 1 to 0 m^{-1} at 0.003.

2.1.2 Design Concentrations of Carbon Monoxide (CO) and Nitrogen Dioxide (NO_2)

The design concentration of CO and NO_2 in the tunnel shall conform to the following provisions:

(1) Under normal traffic conditions, the design concentration of CO in the

tunnel can be taken as shown in Table 2-3.

Design Concentration of CO (δ_{co}) Table 2-3

Tunnel length(m)	≤1000	≥3000
δ_{co} (cm³/m³)	150	100

Note: When the length of the tunnel is 1 000 m < L < 3 000 m, the value can be taken according to the line insertion method.

(2) During traffic block, the average design concentration δ_{co} of CO in the block should be 150 cm³/m³, and the elapsed time should not exceed 20 min.

(3) The average NO_2 concentration δ_{no2} within 20 minutes in the tunnel should be 1.0 cm³/m³.

(4) In the tunnel with mixed traffic of pedestrians and vehicles, the design concentration of CO in the tunnel shall not be greater than 70 cm³/m³, and the design concentration of NO_2 in the tunnel shall not be greater than 0.2 cm³/m³ within 60 minutes.

(5) During the maintenance of the tunnel, the allowable concentration of CO in the air of the tunnel operation section shall not be greater than 30 cm³/m³, and the allowable concentration of NO_2 shall not be greater than 0.12 cm³/m³.

2.1.3 Ventilation Requirements

The ventilation requirements in the tunnel shall comply with the following provisions:

(1) The minimum ventilation frequency in tunnel space shall not be lower than 3 times per hour.

(2) For tunnels with longitudinal ventilation, the ventilation speed shall not be lower than 1.5 m/s.

2.2 Calculation of Air Demand

Guidelines for Design of Ventilation of Highway Tunnels (JTG/T D70/2-02—2014) for air demand calculation are as follows:

(1) The designed hourly traffic volume and the corresponding harmful gas emissions from motor vehicles shall be matched with each design target year.

(2) Benchmark emission of harmful gases from motor vehicles shall be calculated from 2000 to the design target year at a rate of 2% per annum, which shall be taken as the baseline emission of tunnel ventilation design target year, and the maximum reduction period shall not exceed 30 years.

(3) When there is a new type of environmental-friendly engine vehicle in the traffic composition of the road section where the tunnel is located, the emission of harmful gases should be calculated separately.

(4) When determining the air demand, the diluted soot and CO shall be calculated respectively according to the vehicle speed of 10 km/h in each working condition below the design speed of the tunnel, and the air demand for traffic block and ventilation shall be calculated, and the larger one shall be taken as the design air demand.

2.2.1 Air Volume Required For Dilution of Fume

Emissions of fume can be calculated according to Formula (2-1):

$$Q_{VI} = \frac{1}{3.6 \times 10^6} q_{VI} f_{a(VI)} f_d f_{h(VI)} f_{iv(VI)} L \sum_{m=1}^{n_D} [N_m f_{m(VI)}] \quad (2-1)$$

Where: Q_{VI}——Tunnel fume emission (m³/s);

q_{VI}——Design the fume baseline emissions [m³/(veh · km)] for the target year, and the fume baseline emissions for 2000 should be 2.0 m³/(car · km);

$f_{a(VI)}$——Taking into account the smog factor, 1.0 for high-speed and first-class highways and 1.2 to 1.5 for second, third and fourth-class highways;

f_d——Vehicle density coefficient, as per Table 2-4;

$f_{h(VI)}$——Taking into account the altitude coefficient of smog, take the values as shown in Table 2-4, altitude coefficient considering fume are shown in Figure 2-1;

$f_{iv(VI)}$——Longitudinal slope with fume taken into account-the vehicle speed coefficient shall be taken as per Table 2-5;

n_D——Number of diesel vehicle categories;

N_m——Traffic volume of the corresponding models (veh/h);

$f_{m(VI)}$——Type coefficients of diesel vehicles with fume taken into account, as shown in Table 2-6.

Vehicle density coefficient (f_d)   Table 2-4

Operating speed (km/h)	100	80	70	60	50	40	30	20	10
f_d	0.6	0.75	0.85	1	1.2	1.5	2	3	6

Longitudinal slope-speed coefficient considering fume ($f_{iv(VI)}$)   Table 2-5

Operating speed (km/h)	Longitudinal direction of tunnel traffic (i/%)								
	-4	-3	-2	-1	0	1	2	3	4
80	0.30	0.40	0.55	0.80	1.30	2.60	3.70	4.40	—
70	0.30	0.40	0.55	0.80	1.10	1.80	3.10	3.90	—
60	0.30	0.40	0.55	0.75	1.00	1.45	2.20	2.95	3.70
3.7	0.30	0.40	0.55	0.75	1.00	1.45	2.20	2.95	3.70
40	0.30	0.40	0.55	0.70	0.85	1.10	1.45	2.20	2.95
30	0.30	0.40	0.50	0.60	0.72	0.90	1.10	1.45	2.00
10~20	0.30	0.36	0.40	0.50	0.60	0.72	0.85	1.03	1.25

Type coefficient of diesel vehicles considering fume ($f_{m(VI)}$)   Table 2-6

Minibus, light trucks	Medium trucks	Heavy trucks, large buses	Trailer, container trucks
0.4	1.0	1.5	3.0

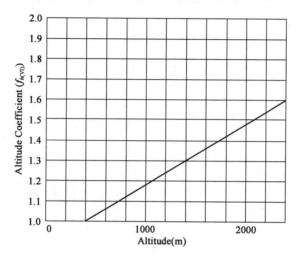

Figure 2-1 Altitude coefficient considering fume ($f_{h(VI)}$)

The air demand for diluting the fume can be calculated according to Formula (2-2):

$$Q_{req(VI)} = \frac{Q_{VI}}{K} \quad (2\text{-}2)$$

Where: $Q_{req(VI)}$——Air requirement for tunnel dilution fume (m³/s);

K——Fume design concentration (m⁻¹).

2.2.2 Air Volume Required for Diluting CO

CO emissions can be calculated according to Formula (2-3):

$$Q_{CO} = \frac{1}{3.6 \times 10^6} q_{CO} f_a f_d f_h f_{iv} L \sum_{m=1}^{n} [N_m f_m] \quad (2\text{-}3)$$

Where: Q_{CO}——Tunnel CO emissions (m³/s);

q_{CO}——CO baseline emissions for the design target year [m³/(veh·km)]: 0.007 m³/(vehicle·km) in 2000 for normal traffic and 0.015 m³/(vehicle·km) for traffic block;

f_d——1.0 for high-speed and first-class highways and 1.1-1.2 for second-class, third-class and fourth-class highways, taking into account the smog condition coefficient;

f_a——Vehicle density coefficient, as shown in Table 2-4;

f_h——Take the altitude coefficient of smog into account and take the value as shown in Figure 2-2;

f_{iv}——Longitudinal Slope Considering Smog-The vehicle speed coefficient shall be calculated according to Table 2-7;

n——Number of diesel vehicle categories;

N_m——The traffic volume of the corresponding models (veh/h);

f_m——Type factor of diesel vehicle considering smog, as shown in Table 2-8.

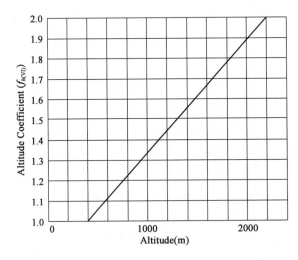

Figure 2-2 Altitude coefficient considering smog ($f_{h(CO)}$)

Longitudinal slope- speed coefficient considering CO (f_{iv})   Table 2-7

Operating speed (km/h)	Longitudinal direction of tunnel traffic ($i/\%$)								
	0.4	0.3	0.2	0.1	0	1	2	3	4
100	1.2	1.2	1.2	1.2	1.2	1.4	1.4	1.4	1.4
80	1.0	1.0	1.0	1.0	1.0	1.0	1.0	1.2	1.2
70	1.0	1.0	1.0	1.0	1.00	1.0	1.0	1.0	1.2
50	1.0	1.0	1.0	1.0	1.0	1.0	1.0	1.0	1.0
40	1.0	1.0	1.0	1.0	1.0	1.0	1.0	1.0	1.0
30	0.8	0.8	0.8	0.8	0.8	1.0	1.0	1.0	1.0
20	0.8	0.8	0.8	0.8	0.8	1.0	1.0	1.0	1.0
10	0.8	0.8	0.8	0.8	0.8	0.8	0.8	0.8	0.8

Vehicle type coefficient considering CO (f_m)   Table 2-8

Vehicle type	Various diesel vehicles	Gasoline vehicle			
		Minibus	Travel car, light trucks	Medium trucks	Large trucks and trailers
f_m	1.0	1.0	2.5	3.2	7.0

The air demand for diluting CO can be calculated according to Formula (2-4):

$$Q_{req(CO)} = \frac{Q_{CO}}{\delta} \frac{p_0}{p} \frac{T}{T_0} \times 10^6 \qquad (2-4)$$

Where: $Q_{req(CO)}$——Air demand for tunnel dilution of CO(m³/s);

δ——CO concentration;
p_0——Standard atmospheric pressure (kN/m^2), 101.325 kN/m^2;
p——Atmospheric pressure at the tunnel site (kN/m^2);
T_0——Standard temperature (K), 237 K;
T——Summer temperature at the tunnel Site (K).

2.2.3 Air Demand for Tunnel Ventilation

The air requirement for tunnel ventilation can be calculate according to Formula (2-5):

$$Q_{req(vent)} = \frac{A_r L n_s}{3600} \tag{2-5}$$

Where: $Q_{req(vent)}$——Tunnel ventilation air demand (m^3/s);
A_r——Tunnel clearance sectional area (m^2);
L——Tunnel length (m);
n_s——Ventilation rates per hour in the tunnel.

In the tunnel with longitudinal ventilation, the air demand for ventilation shall be calculated according to Formula (2-5) and Formula (2-6), and the larger one shall be taken as the air demand for uninterrupted ventilation in the tunnel space, that is:

$$Q_{req(ac)} = v_{ac} A_r \tag{2-6}$$

Where: v_{ac}——Tunnel ventilation speed shall not be lower than 1.5 m/s;
A_r——Tunnel clearance sectional area (m^2).

2.3 Ventilation Methods and Selection

The ventilation methods of highway tunnels can be divided into natural ventilation and mechanical ventilation according to the supply form, air flow state and supply principle, and mechanical ventilation can be divided into longitudinal ventilation, transverse ventilation, semi-transverse ventilation and compound ventilation.

2.3.1 Natural Ventilation

Natural ventilation without ventilation equipment is to use the natural wind pressure between the openings or the traffic ventilation force generated by the action of the automobile driving piston to realize the ventilation and ventilation of the tunnel. In general, natural ventilation may be used in shorter tunnels. For highway tunnels, the following empirical formulas are used to distinguish between natural ventilation and mechanical ventilation:

$$LN \geqslant 6 \times 10^5 \text{ (two-way traffic)} \tag{2-7}$$

Or

$$LN \geqslant 6 \times 10^5 \text{ (one-way traffic)} \tag{2-8}$$

Where: L——Tunnel length(m);
N——Vehicle flow (veh/d).

2.3.2 Longitudinal Ventilation

1. Full-jet longitudinal ventilation

Full-jet longitudinal ventilation is a kind of ventilation, which uses jet fan to generate high-speed air flow to promote the front air to flow longitudinally in the tunnel, so that fresh air flows in from one side of the tunnel and polluted air flows out from the other side of the tunnel.

1) Pressure balance in the tunnel

The full jet longitudinal ventilation diagram is shown in Figure 2-3.

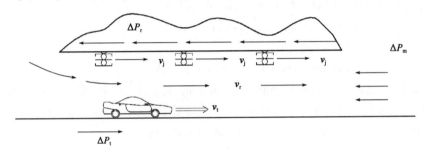

Figure 2-3 Full jet longitudinal ventilation diagram

When the air flow in the tunnel stabilizes, according to the Bernoulli equation, the following results can be obtained:

$$\Delta P = \Delta P_r + \Delta P_m - \Delta P_t \tag{2-9}$$

Where: ΔP——Ventilation pressure provided by the jet fan(N/m^2);

ΔP_r——Tunnel friction and local resistance loss at the entrance and exit (N/m^2);

ΔP_m——Wind pressure generated by natural wind(N/m^2);

ΔP_t——Wind pressure generated by traffic wind(N/m^2).

ΔP_r is the sum of the local resistance loss caused by the flow of air into and out of the tunnel opening and the along-path resistance loss caused by the flow of air in the tunnel, the value of which can be calculated according to Formula (2-10):

$$\Delta P_r = \left(\zeta_e + \zeta_0 + \lambda \frac{L}{D_r}\right)\frac{\rho}{2}v_r^2 \tag{2-10}$$

Where: ζ_e——Local resistance coefficient at tunnel entrance, generally 0.6;

ζ_0——Local resistance coefficient at tunnel exit, generally 1;

λ——Frictional resistance coefficient related to the relative roughness of the tunnel lining surface;

L——Tunnel length(m);

D_r——Equivalent diameter of tunnel clearance section(m), $D_r = \dfrac{4A_r}{C_r}$;

A_r——Section area of tunnel clearance(m^2);
C_r——Circumference of tunnel section(m);
v_r——Designed wind speed in tunnel(m/s), $v_r = Q_{req}/A_r$;
ρ——Air density.

ΔP_m is the wind pressure generated by natural wind, which generates thrust when it is in the same direction as the traffic wind and resistance when it is in the opposite direction. Its value can be calculated according to Formula (2-11):

$$\Delta P_m = \left(\zeta_e + \zeta_0 + \lambda \frac{L}{D_r}\right)\frac{\rho}{2}v_n^2 \qquad (2\text{-}11)$$

Where: v_n—— Wind speed in tunnels due to natural wind action(m/s).

The wind pressure generated by the traffic wind of a single-hole two-way traffic tunnel can be calculated according to Formula (2-12):

$$\Delta P_t = \frac{A_m}{A_r}\frac{\rho}{2}n_+(v_{t(+)} - v_r)^2 - \frac{A_m}{A_r}\frac{\rho}{2}n_-(v_{t(-)} - v_r)^2 \qquad (2\text{-}12)$$

Where: A_m——Equivalent impedance area of the vehicle (m^2);
$\quad n_+$——Number of vehicles in the tunnel in the same direction as v_r (veh),
$$n_+ = \frac{N_+ L}{3600 v_{t(+)}};$$
$\quad n_-$——Number of vehicles in the tunnel opposite to v_r(veh), $n_- = \frac{N_- L}{3600 v_{t(-)}}$;
$\quad N_+$——Designed peak hour traffic in the same direction as v_r in the tunnel(veh/h);
$\quad N_-$——Designed peak hour traffic opposite to v_r in the tunnel(veh/h);
$\quad v_r$——Designed wind speed of the tunnel(m/s), $v_r = \frac{Q_r}{A}$;
$\quad Q_r$——Designed air volume of the tunnel(m^3/s);
$\quad v_{t(+)}$——Vehicle speeds in all conditions in the same direction as v_r(m/s);
$\quad v_{t(-)}$——Vehicle speeds in all conditions opposite to v_r(m/s).

The wind pressure ΔP_t generated by traffic wind in one-way traffic tunnel can be calculated according to Formula (2-13):

$$\Delta P_t = \frac{A_m}{A_r}\frac{\rho}{2}n_C(v_t - v_r)^2 \qquad (2\text{-}13)$$

Where: n_C——Number of vehicles in the tunnel, $n_C = \frac{NL}{3600 v_t}$;
$\quad v_t$——Vehicle speed under various working conditions (m/s).

2) Calculation of jet fan lifting pressure and number of jet fans required

The lifting pressure of each jet fan is calculated according to Formula (2-14):

$$\Delta P_j = \rho v_j^2 \frac{A_j}{A_r}\left(1 - \frac{v_r}{v_j}\right)\eta \qquad (2\text{-}14)$$

Where: ΔP_j——Lifting pressure of a single jet fan (N/m^2);
v_j——Wind speed of the jet fan(m/s);
v_r——Designed wind speed in tunnel(m/s);
A_j——Area of air outlet of jet fan (m^2);
A_r——Cross-sectional area of tunnel clearance (m^2);
η——Friction loss reduction factor of jet fan position, when one jet fan is arranged on the same section of the tunnel, it can be valued according to Table 2-9. When two or more jet fans are arranged on the same section of the tunnel, 0.7 shall be taken.

Friction Loss Reduction Coefficient of Single Jet Fan   Table 2-9

$\dfrac{Z}{D_j}$	1.5	1.0	0.7	diagram
η	0.91	0.87	0.85	

Where: Z——Fan center distance from tunnel wall(mm);
D_j——Fan outlet diameter(mm).

Calculate the number of jet fans according to the Formula (2-15):

$$i = \frac{\Delta P_\tau + \Delta P_m - \Delta P_t}{\Delta P_j} \qquad (2\text{-}15)$$

Where: i——Number of jet fans;
ΔP_j——Lifting pressure of a single jet fan (N/m^2);
ΔP_τ——Tunnel friction and local resistance loss at the entrance and exit (N/m^2);
ΔP_m——Wind pressure generated by natural wind(N/m^2);
ΔP_t——Wind pressure generated by traffic wind(N/m^2).

2. Vent shaft drainage vertical ventilation

The ventilation facilities of the exhaust longitudinal ventilation of the ventilation shaft are composed of a vertical shaft, an air duct and a fan. When the tunnel is a one-way traffic tunnel, the vertical shaft should be located at the exit side of the tunnel. When the tunnel is a two-way traffic tunnel, the shaft should be located in the middle of the longitudinal length of the tunnel. The working method of the fan is exhaust ventilation. Fresh air enters the tunnel through the openings on both sides and polluted air is discharged from the tunnel through the vertical shaft. The vertical ventilation by ventilation shaft is adopted, and the highest concentration of harmful gases in the tunnel is in the vertical shaft. The detection of harmful gases should be strengthened here. Ventilation shaft discharge type can be changed into ventilation shaft feed type, as long as the fan's working method is changed from exhaust ventilation to supply type.

1) Design of ventilation shaft exhaust longitudinal ventilation in two-way traffic tunnel

Figure 2-4 shows the pressure pattern of the two-way traffic tunnel vent shaft exhaust longitudinal ventilation.

Chapter 2 Operation Ventilation of Highway Tunnels

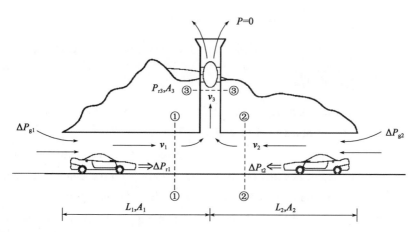

Figure 2-4 Pressure Pattern Diagram of Shaft Exhaust Longitudinal Ventilation in Bi-directional Traffic Tunnels

Where: ΔP_{g1} is the meteorological pressure difference between the tunnel portal and the ventilation shaft outlet in Section 1 (N/m^2), it is positive when the natural wind is in the same direction as the tunnel ventilation;

ΔP_{g2} is the meteorological pressure difference between the tunnel portal and the ventilation shaft outlet in Section 2 (N/m^2), it is positive when the natural wind is in the same direction as the tunnel ventilation;

L_1 is the length of Section 1, (m);

L_2 is the length of Section 2, (m);

v_1 is the average wind speed of the cross section ①-①in Section 1 (m/s);

v_2 is the average wind speed of the cross section ②-②in Section 2 (m/s);

v_3 is the average wind speed of the cross section ③-③in Section 3 (m/s);

A_1 is the area of cross section ①-①in Section 1 (m^2);

A_2 is the area of cross section ②-②in Section 2 (m^2);

A_3 is the area of cross section ③-③in Section 3 (m^2);

P_{r3} is the static pressure of cross section ③-③in Section 3 (Pa);

ΔP_{t1} is the traffic wind pressure of Section 1 (N/m^2);

ΔP_{t2} is the traffic wind pressure of Section 2 (N/m^2).

The air pressure required for longitudinal ventilation of centralized exhaust air in bi-directional traffic tunnels is as follows:

$$\Delta P = \Delta P_0 + \Delta P_s \qquad (2\text{-}16)$$

Where: ΔP_0——The pressure difference between the air at the tunnel entrance and the air in the tunnel at the bottom of the ventilation shaft (N/m^2);

ΔP_s——Friction resistance of shaft and the loss at entrance and exit (N/m^2).

ΔP_0's value is calculated according to Formula (2-17):

$$\Delta P_0 = \Delta P_\tau \pm \Delta P_t \pm \Delta P_m \qquad (2\text{-}17)$$

Where: ΔP_τ——The sum of the tunnel friction and the local loss of the entrance (N/m^2), $\Delta P_\tau = \left(\zeta_e + \lambda \dfrac{L}{D_r}\right)\dfrac{\rho}{2}v_r^2$;

ΔP_t——Wind pressure generated by traffic wind (N/m^2), the specific calculation is as shown in Formulas (2-12) and (2-13);

ΔP_m——The equivalent pressure difference between the tunnel entrance and the pressure reference point is generally 10 Pa when there is no measured data, and one end of the tunnel entrance is generally taken as the reference point.

The above three parts are positive or negative depending on whether they are favorable for ventilation.

The tunnel sections on the left and right sides of the vertical shaft are calculated as ΔP_0, and the larger values of ΔP_0 and ΔP_0 are taken as the design values.

The value of ΔP_s is calculated according to Formula (2-18):

$$\Delta P_s = \left(\zeta_s + \zeta_0 + \lambda_s \dfrac{L_s}{D_s}\right)\dfrac{\rho}{2}v_s^2 \qquad (2\text{-}18)$$

Where: ζ_s——Convergence and bending loss coefficient;

ζ_0——Local resistance coefficient of shaft outlet;

λ_s——Coefficient of frictional resistance related to the relative roughness of the shaft surface;

L_s——Vertical shaft height (m);

D_s——Equivalent diameter of that clearance section of the vertical shaft (m), $D_s = \dfrac{4A_s}{C_s}$;

A_s——Area of clearance section of vertical shaft (m^2);

C_s——Perimeter of vertical shaft section (m);

v_s——Designed wind speed in the vertical shaft (m/s);

ρ——Air density (kg/m^3).

2) Design of ventilation shaft drainage vertical ventilation in unidirectional traffic tunnel

When there are strict environmental requirements near the exit of one-way traffic, that is, the pollution wind in the tunnel is not allowed to blow out of the tunnel (exit), it is advisable to adopt the ventilation shaft discharge type longitudinal ventilation.

The pressure pattern of the split vent shaft discharge longitudinal ventilation of the unidirectional traffic tunnel can be seen in Figure 2-5.

The calculation principle of air pressure required for longitudinal ventilation of centralized exhaust air in unidirectional traffic tunnel is consistent with that of bidirectional traffic tunnel, and will not be described here.

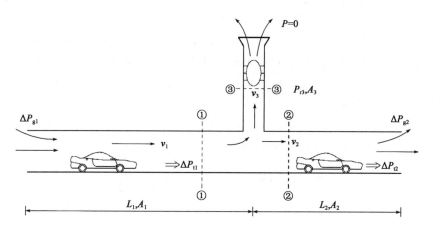

Figure 2-5 Pressure pattern diagram of shaft exhaust longitudinal ventilation in unidirectional traffic tunnels

3. Ventilation vertical shaft supply and exhaust longitudinal ventilation

The ventilation shaft supply and exhaust longitudinal ventilation is provided with a supply shaft and an exhaust shaft, and the polluted air in the tunnel is discharged from the exhaust shaft, and the fresh air enters the tunnel from the supply shaft. This ventilation can effectively utilize the traffic ventilation pressure and is suitable for long highway tunnels with unidirectional traffic. It can also be used for tunnels with two-way traffic in the near future and one-way traffic in the long term. The ventilation method is shown in Figure 2-6.

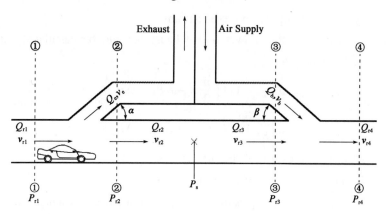

Figure 2-6 Shaft Supply and Exhaust Longitudinal Ventilation Diagram

1) Calculation of supply and exhaust port lifting pressure

The momentum equation is established along the longitudinal axis of the tunnel as follows:

$$A(P_{r1} - P_{r2}) = \rho Q_s v_{r2} + \rho Q_e v_{e2} \cos\alpha - \rho Q_{r1} v_{r2} \quad (2\text{-}19)$$

$$A(P_{r3} - P_{r4}) = \rho Q_{r4} v_{r4} + \rho Q_b v_b \cos\beta - \rho Q_s v_{r3} \quad (2\text{-}20)$$

Where: A —— Cross-sectional area of tunnel (m^2);

ρ —— Air density (kg/m^3);

$P_{r1}, P_{r2}, P_{r3}, P_{r4}$ —— Static pressure of section 1, 2, 3, 4 (N/m^2);

$v_{r1}, v_{r2}, v_{r3}, v_{r4}$ —— Wind speed of section 1, 2, 3, 4 (m/s);

Q_{r1}, Q_{r4}, Q_s ——Air volume in sections 1, 4 and short lanes (m³/s);

v_e, Q_e ——Wind speed(m/s) and air volume(m³/s) of the exhaust port;

v_b, Q_b ——Wind speed(m/s) and air volume(m³/s) of the supply port;

α, β ——The angle between exhaust duct, outside section of supply duct and tunnel (°).

According to the continuity equation, $Q_s = Q_{r1} - Q_e$ is obtained, so that $v_{r2} = Q_s/A = v_{r1}(1 - Q_e/Q_{r1})$, substitution (2-19) is obtained.

$$P_{r1} - P_{r2} = 2\frac{Q_e}{Q_{r1}}\left(\frac{Q_e}{Q_{r1}} - 2 + \frac{v_e}{v_{r1}}\cos\alpha\right)\frac{\rho v_{r1}^2}{2} \quad (2-21)$$

By the same token:

$$P_{r3} - P_{r4} = 2\frac{Q_b}{Q_{r4}}\left(2 - \frac{Q_b}{Q_{r4}} - \frac{v_b}{v_{r4}}\cos\beta\right)\frac{\rho v_{r4}^2}{2} \quad (2-22)$$

Let $P_{r2} - P_{r2} = \Delta P_e$, $P_{r4} - P_{r3} = \Delta P_b$, referred to as the lifting pressure of the exhaust port and the supply port, respectively, be substituted into Formulas (2-21) and (2-22) to obtain the lifting pressure of the exhaust port and the supply port, respectively:

$$\Delta P_e = 2\frac{Q_e}{Q_{r1}}\left(2 - \frac{v_e}{v_{r1}}\cos\alpha - \frac{Q_e}{Q_{r1}}\right)\frac{\rho v_{r1}^2}{2} \quad (2-23)$$

$$\Delta P_b = 2\frac{Q_b}{Q_{r4}}\left(\frac{Q_b}{Q_{r4}} + \frac{v_b}{v_{r4}}\cos\beta - 2\right)\frac{\rho v_{r4}^2}{2} \quad (2-24)$$

2) Design air pressure of supply and exhaust fans

The design air pressure of the blower and exhaust fan can be calculated according to the Formula (2-25) and Formula (2-26):

$$\Delta P_{totb} = 1.1\left(\frac{\rho}{2}v_b^2 + \Delta P_{sb} + \Delta P_b\right) \quad (2-25)$$

$$\Delta P_{tote} = 1.1\left(\frac{\rho}{2}v_e^2 + \Delta P_{se} + \Delta P_e\right) \quad (2-26)$$

Where: ΔP_{totb} ——Design air pressure of the blower (N/m²);

ΔP_{tote} ——Design air pressure of exhaust fan (N/m²);

ΔP_{sb} ——The sum of the along-path resistance and the local resistance from the ventilation shaft supply port to the supply port in the tunnel;

ΔP_{se} ——The sum of the along-path resistance and the local resistance from the exhaust port to the exhaust port of the ventilation shaft;

ρ ——Air density (kg/m³).

2.3.3 Transverse Ventilation

The transverse ventilation is that the fresh supply duct and the polluted air exhaust duct are arranged in the tunnel, and only the air flow in the transverse direction is allowed in the tunnel, and the longitudinal air flow is not generated basically, as shown in Figure 2-7. In two-way traffic, the longitudinal wind speed of the lane is approximately zero, and the pollutant concentration distribution is generally uniform along the whole length of the tunnel. However, in one-way traffic, because of the influence of traffic wind caused by vehicle driving, a certain wind speed can be generated in the longitudinal direction, and the pollutant concentration increases gradually from the inlet to the outlet, but most of the polluted air is still discharged through the exhaust duct. The air flow of transverse ventilation is circulating in the cross-section of the tunnel, and the wind speed in the lane is low, and the fume exhausting effect is good, which is especially suitable for the long tunnel with two-way traffic.

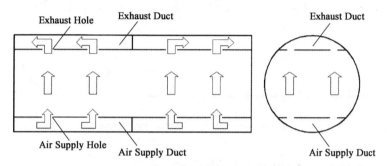

Figure 2-7 Transverse ventilation Schematics

The supply system of that full-transverse ventilation and the supply ventilation semi-transverse ventilation are generally sucked in fresh air by the supply tower, pressurized by the press-in ventilator, then the air is sent into the supply channel of the tunnel through the connecting air duct, and the air is sent into the lane space through the supply port. The design full pressure of the blower ΔP_{totb} can be calculated according to the formula (2-27):

$\Delta P_{totb} = 1.1 \times$ (Tunnel air pressure + Required end pressure of the supply duct + Static pressure difference of the supply duct + Dynamic pressure at the beginning of the supply duct + Pressure loss of the connecting duct) (2-27)

The exhaust system of full transverse ventilation and exhaust semi-transverse ventilation is to discharge the polluted air from the lane space through the exhaust port, exhaust duct and connecting duct, and to discharge the polluted air from the tunnel through the exhaust tower by the exhaust fan plus negative pressure. The design full pressure ΔP_{tote} of the exhaust fan can be calculated according to Formula (2-28):

$\Delta P_{totb} = 1.1 \times$ (Required start-end pressure of exhaust duct + Static pressure difference of exhaust duct - Dynamic pressure at the end of exhaust duct + Pressure loss of connecting duct) (2-28)

2.3.4 Semi-transverse Ventilation

The semi-transverse ventilation is that the fresh supply duct is arranged in the tunnel, and the fresh air is mixed with the polluted air in the driveway and then flows longitudinally along the tunnel to the openings at both ends of the tunnel for discharge, as shown in Figure 2-8. The ventilation is characterized by transverse uniform direct air intake and direct dilution of automobile exhaust, which is beneficial to the follow-up vehicles. If there are pedestrians, pedestrians can directly inhale fresh air. Semi-transverse ventilation is a kind of ventilation between longitudinal and transverse ventilation, which integrates the advantages and disadvantages of longitudinal and transverse ventilation. In some long tunnels, semi-transverse ventilation can be considered because of full transverse ventilation's high cost.

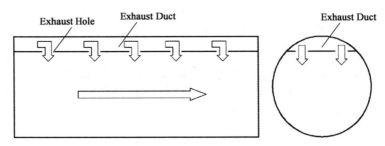

Figure 2-8　Semi-transversal Ventilation Schematics

2.3.5　Selection of Ventilation Methods

The main factors influencing the selection of ventilation methods are:

(1) **Tunnel length**. When the traffic volume is constant, the longer the tunnel is, the more exhaust gas accumulates in the tunnel, and the more air volume is needed in the design. At the same time, the longer the tunnel, the greater the loss caused by accidents and disasters, and the higher the ventilation safety and reliability requirements.

(2) **Tunnel traffic conditions**. Tunnel traffic condition means that the tunnel is one-way or two-way traffic and the traffic volume of the tunnel. One-way road tunnels can make full use of natural air and piston air and are suitable for longitudinal or semi-transverse ventilation. Tunnels with high traffic volume have high concentrations of harmful gases, which are suitable for transverse ventilation or semi-transverse ventilation.

(3) **Geological conditions**. If the tunnel is located in a good geological condition, the construction cost is lower, then we can choose the higher cost of transverse or semi-transverse ventilation. Conversely, if the tunnel is located in poor geological conditions, the construction cost is higher, then the choice of transverse or semi-transverse ventilation will be affected.

(4) **Topographic and meteorological conditions**. The topography and meteorological conditions of the tunnel affect the direction and flow of natural wind in the tunnel. When the natural air flow is relatively large and the flow direction is relatively stable, the short tunnel can be ventilated directly. If the natural air flow changes greatly and has great influence on the longitudinal ventilation effect, the transverse or semi-transverse ventilation can be selected.

Table 2-10 and 2-11 list the advantages and disadvantages of various ventilation methods, which can be used as a reference in the selection of ventilation methods.

Chapter 2 Operation Ventilation of Highway Tunnels

Characteristics of various ventilation methods (two-way traffic tunnel)   Table 2-10

Ventilation method		Longitudinal			Semi-transverse		Transverse
Basic characteristics		The airflow of ventilation flow along that longitudinal direction of the tunnel			The air is supply or exhausted by that tunnel air duct, and the air is exhaust or exhausted along the longitudinal direction of the tunnel by the opening of the tunnel		The supply and exhaust duct are respectively arranged, and that ventilation air flow is longitudinally flow in the tunnel
Form of representation		Full jet	Centralized supply at tunnel opening	Exhaust form of ventilation shaft	Supply of semi-transverse ventilation	Exhaust semi-transverse ventilation	
Formal characteristics		Lifting pressure by the group of jet jan	Lifting pressure by jet supply	Two-end intake and central exhaust	Supply by supply duct	Ventilation by exhaust duct	
General character	Suitable length for non-fire conditions	1500-3000m	About 1500 m	About 4000 m	About 3000 m	About 3000 m	Unrestricted
	Utilization of traffic wind	Not good	Not good	Very good	Good	Not good	Not good
	Noise	Comparatively high	Comparatively high	Comparatively low	Low	Low	Low
	Fume exhaust in fire	Inconvenient	Comparatively convenient	Comparatively convenient	Convenient	Convenient	Good effect
	Construction cost	Low	Common	Common	Comparatively high	Comparatively high	High
	Management and maintenance	Inconvenient	Convenient	Convenient	Common	Common	Common
	Implementation by stages	Easy	Not easy	Not easy	Difficult	Difficult	Difficult
	Technical difficulty	Not difficult	Common	Common	Slightly difficult	Slightly difficult	Difficult
	Operation expense	Low	Common	Common	Comparatively high	Comparatively high	High
	Environmental protection at opening	Adverse	Adverse	Favorable	Common	Favorable	Favorable

When selecting ventilation methods, the tunnel length, radius of flat curve, longitudinal slope, elevation engineering, traffic conditions, geological and topographic conditions and meteorological conditions should be considered comprehensively. Reasonable ventilation is a safe and reliable, easy to build and install, less investment, tunnel internal environment is good, the ability to adapt to disasters, easy operation and maintenance of ventilation. However, each ventilation method has its own advantages and disadvantages, so the practical rationality is to realize economy and convenience as far as possible under the premise of ensuring safety and reliability.

Characteristics of various ventilation methods (one-way traffic tunnel)   Table 2-11

Ventilation		Longitudinal				Semi-transverse		Transverse
Fundamental characteristics		Flow of ventilation flows longitudinally along the tunnel				Air is blown or exhausted by the tunnel air duct, and the wind is drawn from the tunnel opening along the longitudinal direction of the tunnel.		Flow of ventilation flows longitudinally along the tunnel because of supply and exhaust ducts
Representative form		Full jet	Centralized-supply at tunnel entrance	Exhaust form of ventilation shaft	Sending and exhausting of ventilation shaft	Supply of semi-transverse ventilation	Exhaust semi-transverse ventilation	
Formal characteristics		Lifting pressure by the group of jet jan	Lifting pressure by jet supply	Two-end intake and central exhaust	Lifting pressure by jet supply	Supply by supply duct	Ventilation by exhaust duct	
General character	Suitable length for non-fire conditions	Within 5000 m	About 3000 m	About 5000 m	Unrestricted	3000-5000 m	About 3000 m	Unrestricted
	Utilization of traffic wind	Very good	Very good	Partial good	Very good	Comparatively good	Not good	Not good
	Noise	Comparatively high	Comparatively high	Comparatively low	Comparatively low	Low	Low	Low
	Fume exhaust in fire	Inconvenient	Inconvenient	Slightly convenient	Slightly convenient	Convenient	Convenient	Good effect
	Construction cost	Low	Common	Common	Common	Comparatively high	Comparatively high	High
	Management and maintenance	Inconvenient	Convenient	Convenient	Convenient	Common	Common	Common
	Implementation by stages	Easy	Not easy	Not easy	Not easy	Difficult	Difficult	Difficult
	Technical difficulty	Not difficult	Common	Common	Slightly difficult	Slightly difficult	Slightly difficult	Difficult
	Operation expense	Low	Common	Common	Common	Comparatively high	Comparatively high	High
	Environmental protection at opening	Adverse	Adverse	Adverse	Common	Common	Favorable	Favorable

Chapter 2 Operation Ventilation of Highway Tunnels

Exercise

2.1 Briefly describe the components and hazards of harmful gases in highway tunnels.

2.2 What are the specific requirements for air quality in highway tunnels?

2.3 Road tunnels on a grade I highway. According to the known conditions, the air demand of the tunnel is calculated. Known:

(1) Design speed: 60 km/h;

(2) Lane and driving conditions: a two-lane, one-way, two-lane tunnel with a daily maximum of 60000 vehicles, including 7000 diesel vehicles, 17000 minibuses, 14000 station wagons, light goods vehicles, 13000 medium goods vehicles and 9000 buses;

(3) Tunnel length and longitudinal slope: the length is 1 600 m, and the longitudinal slope is 2%;

(4) Altitude and air pressure: The average altitude is 1 200 m and the air pressure is 90 kN/m^2;

(5) Tunnel clearance area: 65 m^2;

(6) Tunnel design temperature in summer: 33 ℃.

2.4 What are the ventilation methods for highway tunnels? What are their characteristics?

2.5 What are the methods of longitudinal mechanical ventilation? What's the difference?

2.6 For a two-way driving tunnel, the ratio of up and down vehicles is 1:1. The tunnel adopts full jet longitudinal ventilation. According to the known conditions, the number of fans required for the tunnel is solved. Known:

(1) The tunnel length: L = 850 m, the slope gradient: 0.6%;

(2) The cross-sectional area of the tunnel: $A_r = 53$ m^2, and the equivalent diameter: $D_r = 6$ m;

(3) Design peak hour traffic volume: N = 1 600 veh/h;

(4) Automobile equivalent impedance area: $A_m = 2.5$ m^2;

(5) Design speed: $v_t = 14.0$ m/s, natural wind velocity: $v_n = 14.0$ m/s;

(6) Equivalent pressure difference caused by natural wind: $\Delta P_m = 10.95$ Pa;

(7) Tunnel design air demand: Q = 75 m^3/s;

(8) Tunnel site air density: $\rho = 1.224$ kg/m^3;

(9) Local resistance coefficient at tunnel entrance: $\varepsilon_e = 0.6$, local resistance coefficient at tunnel exit: $\varepsilon_0 = 1$, friction resistance coefficient: $\lambda = 0.025$;

(10) Selection of 900 type jet fan, wind speed: $v_j = 25$ m/s, air outlet area: $A_j = 0.636$ m^2, friction loss reduction factor of jet fan position: $\eta = 0.91$;

2.7 What factors should be considered when selecting ventilation methods for highway tunnel operation?

Key vocabulary:

benchmark emission　基准排放
design target year　设计目标年限
lifting pressure of a jet fan
　射流风机升压力
piston wind　活塞风
traffic wind　交通风

blind heading　独头掘进
full-jet longitudinal ventilation　全射流纵向通风
longitudinal ventilation　纵向通风
semi-transverse ventilation　半横向通风
transverse ventilation　横向通风

Chapter 3　Operation Ventilation of Railway Tunnels

[Important and Difficult Contents of this Chapter]
(1) Hygienic standard of railway tunnel operation.
(2) Calculation of air demand for railway tunnel operation ventilation.
(3) Various ventilation methods and their characteristics of the tunnel.
(4) Pressure pattern and calculation method of longitudinal ventilation.
(5) The principle of selecting ventilation method of railway tunnel.

3.1　Hygienic Standard for Operation of Railway Tunnels

Trains in railway tunnels are mainly electric locomotives and diesel locomotives. Ozone and quartz dust are the main harmful substances produced during the running of electric locomotives because of the variety of goods in railway tunnels. Besides quartz powder vehicles there are also animal and plant dust in the tunnels. The hygienic standard of railway tunnel operation ventilation is strictly in accordance with the requirements of *Code for Design on Operation Ventilation of Railway Tunnel* (*TB 10068—2010*), in which the harmful substances produced by diesel locomotives are mainly carbon monoxide and nitrogen oxides. See Table 3-1 for the air hygiene standard of the operation tunnel for electric locomotive traction and Table 3-2 for the air hygiene standard of the operation tunnel for diesel locomotive traction.

Hygienic standard for air in electric locomotive tunnels　　　Table 3-1

Indicator		Maximum allowable value	Remarks
Ozone (mg/m^3)		0.3	$H < 3000$ m
Dust (mg/m^3)	Quartz dust	8	$M_{SiO_2} < 10\%$
		2	$M_{SiO_2} > 10\%$
	Animal and plant dust	3	—

Hygienic standard for air in diesel locomotive tunnels   Table 3-2

Indicator	Maximum allowable value	Remarks
Carbon monoxide (mg/m^3)	30	$H < 2000$ m
	20	2000 m $\leqslant H \leqslant 3000$ m
	15	$H > 3000$ m
Nitrogen oxides(NO_2) (mg/m^3)	5	$H < 3000$ m

Electric locomotive tunnels, should meet not only the air health standards, but also the temperature and humidity environmental standards in order to make the train drivers and passengers comfortable. See Table 3-3 for the temperature and humidity standard of electric locomotive traction operation tunnel.

Standard for temperature and humidity of electric locomotive tunnels   Table 3-3

Indicator	Maximum allowable value	Remarks
Temperature(℃)	28	—
Humidity(%)	80	—

3.2 Air Demand Calculation

According to the *Code for Design on Operation Ventilation of Railway Tunnel* (*TB 10068—2010*), the ventilation volume of railway tunnels can be calculated according to Formula (3-1):

$$Q = K_i \left(1 - \frac{v_m}{v_T}\right) \frac{FL_T}{t_q} \tag{3-1}$$

Where: Q——Ventilation volume of railway tunnel(m^3/s);

K_i——Correction coefficient of piston air, 1.1 for operation tunnel of diesel locomotive traction and 1 for operation tunnel of electric locomotive traction;

v_m——Piston wind speed(m/s);

v_T——Train speed(m/s);

F——Cross-sectional area of the tunnel(m^2);

L_T——Tunnel length(m);

t_q——Fume exhaust time(s).

According to the *Code for Design on Operation Ventilation of Railway Tunnel* (*TB 10068—2010*), when calculating train piston wind, the unsteady flow theory should be adopted for single-track tunnels with length less than 15 km, the steady flow theory should be adopted for single-track tunnels with length greater than 15 km, and the influence of piston wind can be ignored for double-track tunnels.

Calculate the piston wind speed according to the constant flow theory, as shown in formula (3-2):

Chapter 3 Operation Ventilation of Railway Tunnels

$$v_m = v_T \frac{-1 + \sqrt{1 + \left(\frac{\zeta_m}{K_m} - 1\right)\left(1 \pm \frac{\zeta_n v_n^2}{K_m v_T^2}\right)}}{\frac{\zeta_m}{K_m} - 1} \tag{3-2}$$

Where: v_m——Piston wind speed(m/s);

v_T——Train speed(m/s);

v_n——Natural wind speed(m/s);

K_m——Coefficient of piston wind action;

ζ_m——The drag coefficient of the tunnel section other than the annulus,

$$\zeta_m = 1 + \lambda \frac{L_T - l_T}{d} + \zeta;$$

ζ_n——The total resistance coefficient of the tunnel, $\zeta_n = 1 + \lambda \frac{L_T}{d} + \zeta$;

ζ——Coefficient of resistance at tunnel entrance;

λ——friction resistance coefficient of the tunnel;

l_T——Train length(m);

d——Equivalent diameter of tunnel section(m).

The piston wind action coefficient K_m is calculated according to Formula (3-3):

$$K_m = \frac{Nl_T}{(1 - \alpha)^2} \tag{3-3}$$

$$N = \frac{1}{l_T}\left(0.807\alpha^2 - 1.322\alpha + 1.008 + \lambda_h \frac{l_t}{d_h}\right) \tag{3-4}$$

$$d_h = 4\frac{F - f_T}{S + S_T} \tag{3-5}$$

Where: l_T——Train length(m);

α——Blockage ratio, Ratio of train cross-sectional area f_T to tunnel cross-sectional area F;

N——Reverse train resistance coefficient;

f_T——the sectional area of the train (m²);

λ_h——Coefficient of drag along the inverted annular space airflow rack;

d_h——Equivalent diameter of the inverted annulus(m);

S——Wet perimeter of cross section of inverted side tunnel(m);

S_T——Line-side train section perimeter(m).

Calculate the piston wind speed according to the unsteady flow theory, according to Formula (3-6), Formula (3-7):

When $K_m > \zeta_m$:

$$v_m = \frac{-2AC + ACe^{t\sqrt{B^2 - 4AC}}}{C(B + \sqrt{B^2 - 4AC}) - C(B - \sqrt{B^2 - 4AC})e^{t\sqrt{B^2 - 4AC}}} \tag{3-6}$$

Where: $A = \dfrac{K_m v_T^2 \pm \zeta_n v_n^2}{2\left(L_T + \dfrac{\alpha L_T}{1-\alpha}\right)}$; $B = \dfrac{-K_m v_t}{\left(L_T + \dfrac{\alpha L_T}{1-\alpha}\right)}$; $C = \dfrac{K_m - \zeta_m}{2\left(L_T + \dfrac{\alpha L_T}{1-\alpha}\right)}$;

$$t = \dfrac{1}{\sqrt{B^2 - 4AC}} \ln \left[\dfrac{2Cv_m \left(B + \sqrt{B^2 - 4AC}\right) + 4AC}{2Cv_m \left(B - \sqrt{B^2 - 4AC}\right) + 4AC} \right].$$

When $K_m < \zeta_m$:

$$v_m = \dfrac{-2AC + ACe^{t\sqrt{B^2 + 4AC}}}{C\left(B + \sqrt{B^2 + 4AC}\right) - C\left(B - \sqrt{B^2 + 4AC}\right)e^{t\sqrt{B^2 + 4AC}}} \tag{3-7}$$

Where: $A = \dfrac{K_m v_T^2 \pm \zeta_n v_n^2}{2\left(L_T + \dfrac{\alpha L_T}{1-\alpha}\right)}$; $B = \dfrac{-K_m v_t}{\left(L_T + \dfrac{\alpha L_T}{1-\alpha}\right)}$; $C = \dfrac{\zeta_m - K_m}{2\left(L_T + \dfrac{\alpha L_T}{1-\alpha}\right)}$;

$$t = \dfrac{1}{\sqrt{B^2 + 4AC}} \ln \left[\dfrac{2Cv_m \left(B + \sqrt{B^2 + 4AC}\right) + 4AC}{2Cv_m \left(B - \sqrt{B^2 + 4AC}\right) + 4AC} \right].$$

When the natural wind in the tunnel is in the same direction as the train running direction, the positive sign is taken as A in the equation, and the negative sign is taken as the negative sign.

3.3 Ventilation Methods and Selection for Tunnel Operation

The ventilation methods of railway tunnel operation can be divided into natural ventilation and mechanical ventilation according to the supply form, air flow state and supply principle. The mechanical ventilation can be divided into longitudinal ventilation, transverse ventilation and semi-transverse ventilation.

3.3.1 Natural Ventilation

Ventilation equipment is not provided in natural ventilation. The ventilation of tunnel can be realized by using natural wind pressure between openings or driving piston to generate traffic ventilation wind. In general, natural ventilation may be used in shorter tunnels. For railway tunnels, the following empirical formulas are used to distinguish between natural ventilation and mechanical ventilation:

$$LN \geqslant 100 \tag{3-8}$$

Where: L——Tunnel length (km);
N——Train density (fleet/day).

3.3.2 Longitudinal Ventilation

1. Full-jet longitudinal ventilation

Full-jet longitudinal ventilation is a kind of ventilation, which uses jet fan to generate high-speed air flow to promote the front air to flow longitudinally in the tunnel, so that fresh air flows in from one side of the tunnel and polluted air flows out from the other side of the tunnel.

Under normal operation conditions, the pressure balance in the tunnel satisfies the following formula:

$$P_j + P_m = P_n + P_\zeta + P_\lambda \qquad (3\text{-}9)$$

Where: P_j——Jet fan thrust(N/m^2);

P_m——Train piston wind pressure(N/m^2);

P_n——Natural wind pressure between two holes in the tunnel(N/m^2);

P_ζ——Local resistance(N/m^2);

P_λ——Along-path resistance(N/m^2).

The pressure of a single jet fan can be calculated according to Formula (3-7):

$$p_j = \rho v_j^2 \frac{f}{F}\left(1 - \frac{v_e}{v_j}\right)\frac{1}{K_j} \qquad (3\text{-}10)$$

Where: p_j——Jet fan thrust(N/m^2);

v_e——Mean wind speed (m/s) in the tunnel section(m/s);

v_j——Outlet wind speed of jet fan(m/s);

f——Outlet area of single jet fan(m^2);

K_j——The jet loss coefficient considering the influence of tunnel wall friction is related to the distance between the fan and the wall surface, and can be taken as shown in Figure 3-1.

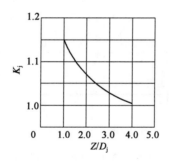

Figure 3-1 Loss coefficient of jet fan

Z-Fan center distance from tunnel wall(mm); D_j-Fan outlet diameter(mm).

So, the number of fans: $i = \dfrac{P_j}{p_j}$

The air pressure of that train piston can be calculated according to Formula (3-11):

$$P_m = K_m \frac{\rho}{2}(v_T - v_m)^2 \qquad (3\text{-}11)$$

The natural wind pressure of the train can be calculated according to Formula (3-12):

$$P_n = \left(\Sigma\zeta + \lambda \frac{L_T}{d}\right)\frac{\rho}{2}v_n^2 \qquad (3\text{-}12)$$

Where: $\Sigma\zeta$——Local resistance coefficient of tunnel entrance and exit;
λ——Coefficient of along-path resistance;
v_n——Natural wind speed (m/s). Natural wind pressure and natural wind action direction is the same, for resistance consideration.

Local resistance can be calculated according to Formula (3-13):

$$P_\zeta = \zeta \frac{\rho}{2} v_e^2 \tag{3-13}$$

Along-path resistance can be calculated according to Formula (3-14):

$$P_\lambda = \lambda \frac{L_T}{d} \frac{\rho}{2} v_e^2 \tag{3-14}$$

2. Segmented longitudinal ventilation

1) exhaust type of converging inclined (vertical) shaft

The facilities of converging inclined (vertical) shaft longitudinal ventilation are composed of inclined (vertical) shaft, air duct and fan. The fan belongs to exhaust ventilation. Fresh air enters the tunnel through the entrance and exit, and the polluted air is discharged from the tunnel through inclined (vertical) shafts. Figure 3-2 shows the pressure pattern of the converging inclined (vertical) shaft longitudinal ventilation

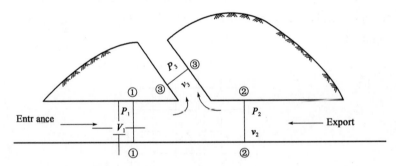

Figure 3-2 Pressure pattern diagram of converging inclined (vertical) shaft longitudinal ventilation

The wind pressure of the cross-section ① and the cross-section; ② inside the tunnel at the bottom of the inclined (vertical) shaft shall meet the Formula (3-15):

$$P_1 + \frac{1}{4}\rho v_1^2 = P_2 + \frac{1}{4}\rho v_2^2 \tag{3-15}$$

Where: $P_1 = P_{n1} - \left(0.5 + \lambda \frac{L_1}{d}\right) \frac{\rho}{2} v_1^2$; $P_2 = P_{n2} - \left(0.5 + \lambda \frac{L_2}{d}\right) \frac{\rho}{2} v_2^2$

The pressure of inclined (vertical) shaft after convergence can be calculated according to Formula (3-16) and Formula (3-17):

$$P_3 = P_{n1} - \left(0.5 + \lambda \frac{L_1}{d}\right) \frac{\rho}{2} v_1^2 - \zeta_{1-3} \frac{\rho}{2} v_3^2 \tag{3-16}$$

$$P_3 = P_{n2} - \left(0.5 + \lambda \frac{L_2}{d}\right) \frac{\rho}{2} v_2^2 - \zeta_{2-3} \frac{\rho}{2} v_3^2 \tag{3-17}$$

Where: P_3——Bottom convergence pressure of inclined (vertical) shaft (N/m^2);

P_{n1}——The natural wind pressure difference between the tunnel entrance and the bottom of the inclined (vertical) shaft (N/m^2);

P_{n2}——The natural wind pressure difference between the exit of the tunnel and the bottom of the inclined (vertical) shaft (N/m^2);

ζ_{1-3}——Local resistance coefficient between section ① and section ③;

ζ_{2-3}——Local resistance coefficient between section② and section③;

v_1——Wind speed of section①(m/s);

v_2——Wind speed of section②(m/s);

v_3——Wind speed of section③(m/s).

The air pressure of the exhaust fan can be calculated according to Formula (3-18):

$$H_g = 1.1\left[P_3 + \frac{\rho}{2}\Sigma\left(\zeta_i + \frac{\lambda_j L_j}{d_j} + 1\right)v_3^2\right] \quad (3\text{-}18)$$

Where: H_g——Exhaust fan wind pressure (N/m^2);

ζ_i——Convergence and bending loss coefficient;

λ_j——Coefficient of friction loss of inclined (vertical) shaft;

L_j——Length of inclined (vertical) shaft (m);

d_j——Equivalent diameter (m) of inclined (vertical) shaft (m).

2) Combined inclined (vertical) shaft supply and exhaust ventilation

A supply shaft and an exhaust shaft are arranged in that longitudinal ventilation of the combined inclined (vertical) shaft to discharge polluted air in the tunnel from the exhaust shaft and send fresh air into the tunnel from the supply shaft. See Figure 3-3 for the pressure pattern of vertical ventilation in inclined (vertical) shafts.

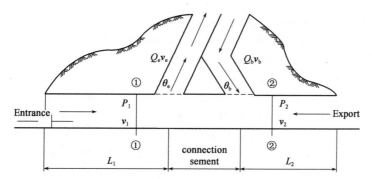

Figure 3-3 Pressure pattern diagram of inclined (vertical) shaft supply and exhaust vertical ventilation

The pressure in the tunnel shall meet the Formula (3-19):

$$P_b + P_e \geqslant P_n + P_\lambda + P_\zeta \quad (3\text{-}19)$$

Where: P_b——Air outlet pressure (N/m^2);

P_e——Exhaust port pressure (N/m^2);

P_n——Natural wind pressure between the tunnel portals (N/m^2);

P_λ——One-way resistance(N/m^2);

P_ζ——Local resistance(N/m^2).

The supply port pressure and the air exhaust port pressure can be calculated according to Formula (3-20) and Formula (3-21):

$$P_b = 2\frac{Q_b}{Q_2}\left(\frac{Q_b}{Q_2} + \frac{v_b}{v_2}\cos\theta_b - 2\right)\frac{\rho v_2^2}{2} \tag{3-20}$$

$$P_e = 2\frac{Q_e}{Q_1}\left(2 - \frac{v_e}{v_1}\cos\theta_e - \frac{Q_e}{Q_1}\right)\frac{\rho v_1^2}{2} \tag{3-21}$$

Where: Q_1——Air volume of section L_1(m^3/s);

v_1——Wind speed of section L_1(m/s);

Q_2——Air volume of section L_2(m^3/s), $Q_2 = Q_b - Q_e + Q_1$;

v_2——Wind speed of section L_2(m/s);

Q_e——Exhaust air volume(m^3/s);

v_e——Wind speed of exhaust port(m/s);

Q_b——Supply air volume;

v_b——Outlet wind speed(m/s).

The air pressure of the supply and exhaust fans can be calculated according to Formula (3-22) and Formula (3-23):

$$H_{gb} = 1.1\left(\frac{\rho}{2}v_b^2 + P_{db} + P_b\right) \tag{3-22}$$

$$H_{ge} = 1.1\left(\frac{\rho}{2}v_e^2 + P_{de} + P_e\right) \tag{3-23}$$

Where: P_{db}——Total pressure loss of supply port, supply shaft and connecting channels(N/m^2);

P_{de}——Total pressure loss of exhaust ports, exhaust shaft and connecting channels(N/m^2).

3.3.3 Transverse Ventilation

Transverse ventilation means that fresh supply duct and polluted air exhaust duct are arranged in the tunnel, only transverse air flow is produced in the tunnel, and longitudinal air flow is not produced basically, as shown in Figure 3-4. Comparing with the longitudinal ventilation, the transverse ventilation produces circulation on the cross-section of the tunnel and carries out air exchange. The wind speed in the driveway is lower and the fume exhausting effect is good. However, it is necessary to set up driveway slab and ceiling and air shaft in the tunnel for transverse ventilation, which makes the tunnel construction work more and more expensive. In addition, due to the restriction of tunnel construction section, small cross-section of supply duct and exhaust duct, large ventilation resistance and large ventilation energy consumption, the operation and management cost is high.

3.3.4 Semi-transverse Ventilation

The semi-transverse ventilation is that the fresh supply duct is arranged in the tunnel, and the fresh air is mixed with the polluted air in the driveway and then flows longitudinally along the tunnel to the openings at both ends of the tunnel for discharge, as shown in Figure 3-5. Semi-transverse ventilation is a kind of ventilation between longitudinal and transverse ventilation, which integrates the advantages and disadvantages of longitudinal and transverse ventilation. In some long tunnels, semi-transverse ventilation may be considered because of full transverse ventilation's high cost.

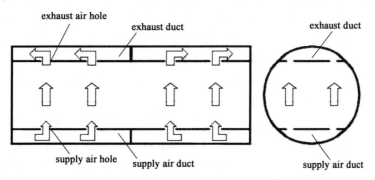

Figure 3-4 Transverse ventilation Schematics

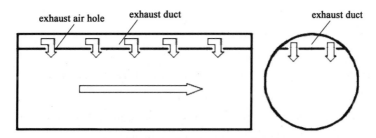

Figure 3-5 Semi-transversal Ventilation Schematics

3.3.5 Selection of Ventilation Methods

The ventilation methods of railway tunnels can be divided into natural ventilation and mechanical ventilation. The selection of ventilation methods should be determined by comprehensive comparison according to technical and economic conditions, safety, effect and other factors. When the tunnel ventilation can be completed by the interaction of train piston wind and natural wind, natural ventilation should be selected. For some extra-long railway tunnels and some tunnels with harmful substances such as gas, mechanical ventilation should be chosen because the combined action of train piston wind and natural wind can not complete tunnel ventilation.

Selection principles of mechanical ventilation:

①Longitudinal ventilation is usually adopted in normal operation;

②When the tunnel is too long or has special requirements, the segmented ventilation can be adopted;

③The combination of fixed ventilation and moving ventilation should be adopted in maintenance operation.

Trains in railway tunnels are mainly electric locomotives and diesel locomotives. Electric locomotives have the characteristics of high speed and low pollution. Therefore, the ventilation of most electrified railway tunnels mainly depends on the piston wind and natural wind generated by the train running in the tunnels, which can meet the ventilation and air sanitation standards of the tunnels without special mechanical ventilation. In *Code for Design on Operation Ventilation of Railway Tunnel* (*TB 10068—2010*), it is stipulated that the tunnel length of electric locomotive traction and passenger dedicated line shall be greater than 20 km, and the tunnel length of passenger and freight co-line shall be greater than 15 km, and mechanical ventilation shall be provided.

Compared with electric locomotive traction tunnel, diesel locomotive traction tunnel has higher concentrations of carbon monoxide, nitrogen dioxide and fume, and more mechanical ventilation. It is stipulated in *Code for Design on Operation Ventilation of Railway Tunnel* (*TB 10068—2010*) that mechanical ventilation should be provided for diesel locomotive traction and tunnel length over 2 km.

Exercise

3.1 Describe the main types of trains running in railway tunnels and the harmful substances produced during the running of different types of trains.

3.2 What are the specific requirements of *Code for Design on Operation Ventilation of Railway Tunnel* (*TB 10068—2010*) for air quality in railway tunnels?

3.3 The total length of a single-track tunnel with electric traction is 15100 m, which is ventilated by full jet longitudinal ventilation. The tunnel section area $F = 31.97$ m^2, equivalent diameter $d = 6.06$ m, wet section circumference $S = 21.1$ m. Ballastless track is adopted, and the wall coefficient of the tunnel is $\lambda = 0.02$. Air density $\rho = 1.225$ kg/m^3, natural backwind speed $v_n = 2.0$ m/s, ventilation in skylight, ventilation time 90 min. The calculated train length is 500 m, the train speed is 80 km/h, the cross-sectional area of the train is 12.6 m^2, the wet circumference is 14.3 m, and the annular resistance coefficient is 0.02. Type 112 jet fan was selected, the wind speed $v_j = 33.9$ m/s, the area of air outlet $A_j = 0.985$ m^2, and the friction loss reduction factor of jet fan position was 0.86. Solve the number of fans required for the tunnel.

3.4 What are the ventilation methods for railway tunnel operation? What are their characteristics?

3.5 Under what conditions must mechanical ventilation be provided for electric locomotive traction tunnels and diesel locomotive traction tunnels? What are the principles for selecting mechanical ventilation?

Key vocabulary:

blockage ratio　　阻塞比
constant flow theory　　恒定流理论

combined inclined (vertical) shaft supply and exhaust ventilation　合斜(竖)井送排式
equivalent diameter　当量直径
exhaust type of converging inclined (vertical) shaft　合流型斜(竖)井排出式
segmented longitudinal ventilation　分段式纵向通风
wetted perimeter　湿周

Chapter 4 Metro Ventilation and Air Conditioning

[**Important and Difficult Contents of this Chapter**]

(1) Composition of ventilation and air conditioning system of metro.

(2) Metro ventilation and air conditioning system and the advantages and disadvantages of each ventilation and air conditioning system.

(3) The internal ventilation system of metro station.

(4) Operation status of ventilation and air conditioning system of metro.

(5) Load calculation method of ventilation and air conditioning system in metro.

4.1 Overview of Metro Ventilation and Air-conditioning System

January 10, 1863, London, the world's first metro line opened to traffic, "Metropolis" because of steam-driven operation, locomotive emissions of fume caused by the underground station environment is difficult to block the hot and humid; After the Metropolis, electric locomotives were introduced into the London Underground, and new problems were encountered. Because of the high power of electric locomotives, more heat was released. With the increase of passenger traffic, the internal environment of the London Underground Station deteriorated further.

In October 1905, when New York's first metro was opened, designers did not consider supply ventilation of tunnels and stations. They believed that vents on sidewalks would provide enough fresh air for the metro system. The next summer, the problem of excessive temperature in the metro became serious because of the poor ventilation on the ground. Later, in order to increase the ventilation rate, more ventilation openings had to be installed on the roof of the station, and fans and ventilation ducts had to be installed in and between the stations.

Drawing lessons from the design of the New York Metro, When the Boston Metro was built in May 1909, Designers are fully aware of the need to provide a comfortable environment for passengers, It is the first time to adopt the air duct at the top of the tunnel to ventilate and increase the area of the station entrance and exit, and it is put forward that "pure air can be obtained by mechanical ventilation". It is concluded that "the temperature problem is related to ventilation, and the temperature difference between the inside and outside of the tunnel will be reduced by increasing the frequency of ventilation and ventilation". Through engineering practice, the internal environment of the metro will be greatly improved.

Chicago's first metro was completed in 1943. From the very beginning of the design of the Chicago Metro, designers were concerned about the environmental control of the station. Edcson Brock made a great contribution to the construction of the ventilation system of this metro. Brock established a method and a formula for calculating the piston effect of trains in *Progress in Ventilation Calculation of Chicago Metro*, In order to achieve heat balance in the metro, Brock considers not only the amount of air change required to maintain a comfortable metro environment. At that same time, the influence of diurnal and annual variation of tunnel wall and soil temperature, The cumulative effects of various temperatures and cycles were measured, and these data were fully utilized in the design of the Chicago Metro, creating a station environmental control system that provides adequate ventilation and a pleasant ambient temperature in the underground station almost all year round without air conditioning.

The successful solution of the environmental problems in the Chicago Metro has led many other cities planning to build metros to start looking for solutions to environmental problems in the early stages of design. The Toronto Metro, which opened in 1954, is based largely on the Chicago Metro. In order to reduce the cost of the project, the distance between the ventilation shafts was nearly tripled by the designers. The blocking ratio of the train is increased by 15%. The piston wind caused by the high-speed train in the tunnel has a lot of negative effects on the physiology and psychology of platform passengers. Subsequently, in order to overcome these adverse effects, Toronto Metro adopted some structural changes and thermal storage (cold) performance of the rock around the tunnel, using night ventilation, to meet better environmental requirements.

Since the first metro was built in London in 1863, nearly 100 large cities in the world now have metros. With the continuous expansion of the scale of urbanization in China, urban population circulation has increased rapidly, traffic congestion has become increasingly serious, the traditional means of public transport has been unable to meet the daily travel needs of the urban population. Metro has fast, convenient, environmental protection, large passenger traffic characteristics, so it becomes an effective means of transportation to solve the traffic tension in modernization cities. The first metro line in China was built in Beijing in July 1965 and began trial operation in January 1971. Subsequently, Shanghai Metro, Guangzhou Metro, Shenzhen Metro and Nanjing Metro were successively built and put into operation, then Hangzhou Metro, Shenyang Metro, Xi'an Metro and so on are under construction. With the operation of Metro, Metro ventilation and air conditioning system (abbreviated as environmental control system) has become a key process system to meet and ensure the internal air environment for the operation of personnel and equipment, and is an indispensable part of metro.

4.2 Composition of Metro Ventilation and Air-conditioning System

4.2.1 Metro Ventilation and Air Conditioning System

The purpose of environmental control system of urban rail transit is to provide comfortable environment for metro passengers during normal operation, and to help passengers leave dangerous places quickly and reduce losses as much as possible in case of emergency. The environmental control system of an urban rail transit line must meet the following three basic requirements.

(1) When the train is running normally, the environmental control system can control the air temperature, humidity, velocity, cleanliness, air pressure and noise in the urban rail transit system reasonably and effectively according to the seasonal climate, so as to provide a comfortable and hygienic air-conditioning environment.

(2) When the train is blocked, the environmental control system can ensure the air circulation in the tunnel, the train air conditioner runs normally and the passengers feel comfortable.

(3) In case of emergency, the environmental control system can control the direction of fume, heat and gas diffusion, and provide safety guarantee for passenger evacuation and rescue personnel entry.

According to the ventilation form of urban rail transit tunnel and the separation relationship between tunnel and station platform, urban rail transit ventilation and air conditioning system is generally divided into three types: open system, closed system and screen door system.

1. Open system

The inside of the tunnel communicates with the outside atmosphere, and only piston ventilation or mechanical ventilation is considered. It is a way of ventilation and ventilation by connecting the piston air shaft, station entrance and exit, and both ends of the tunnel with outdoor air, as shown in Figure 4-1. It is mainly used in the north. In China, the system is used in Beijing Metro Line 1 and Ring Road.

2. Closed system

The closed system is a kind of way that the air in the underground station is not connected with the outdoor air basically, that is, all the ventilation shafts and air doors connected with the outdoor in the urban rail transit station are closed, air conditioning is adopted in the station in summer, and only the minimum fresh air volume or fresh air of air conditioning is provided to the station from the outdoor through the fan. The air-conditioned air of the station is carried into the section through the piston effect of the train, so that the temperature in the section is cooled, and circuitous air passages are arranged at both ends of the station to meet the pressure relief requirements of the piston air in closed operation, and the openings exposed to the ground of the line are isolated by an air curtain to prevent heat and humidity exchange between the openings. The closed system is controlled by wind-jet and can be operated on or off. There are Guangzhou Metro Line 1, Shanghai Metro Line 2, Nanjing Metro Line 1 and Harbin Metro Line 1.

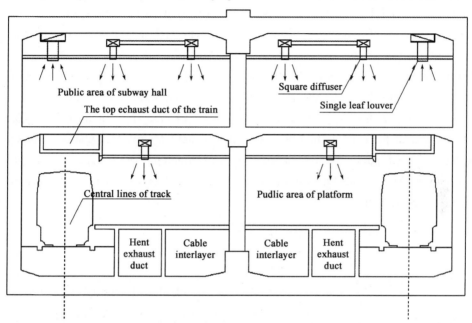

Figure 4-1 Island station exhaust system

There is another closed system, the large surface cooler closed system, The air treatment method is basically the same as that of the above-mentioned closed system, except that the tunnel accident fan is multifunctional to replace the centrifugal fan and the return and exhaust fan of the combined air conditioning unit, and the structural air conditioning equipment, the air filtering device and the fin heat exchanger are arranged in the air duct of the civil structure. There are Nanjing Metro Line 2, Beijing Metro Line 4, Beijing Metro Line 5, Beijing Metro Line 10, Fuzhou Metro Line 8.

In the closed system of urban rail transit, in order to increase the safety of passengers, many stations have set up safety doors at the edge of the platform, but they do not isolate the tunnel from the station air.

3. Screen door system

The screen door is installed at the edge of the platform. It is a transparent barrier with doors built at the edge of the platform. The common area of the platform is completely shielded from the tunnel track area. The spacing distance between the movable doors on each door of the screen door corresponds to the distance between the doors on the train. It looks like the doors of a row of elevators, as shown in Figure 4-2. When the train arrives at the station, the train door faces the movable door on the screen door, and passengers can get on and off the train freely. After the screen door is closed, a partition wall formed can effectively prevent heat flow, air pressure fluctuation and dust in the tunnel from entering the station, effectively reduce the air conditioning load, and create a more comfortable environment for the station. In addition, the screen door system can effectively prevent passengers from falling into the track intentionally or unintentionally, reduce the influence of noise and piston wind on platform waiting passengers. It improves the comfort of passenger waiting environment and lays a technical foundation for unmanned rail transit. However, the initial investment cost of screen door is high, and the reliability of train parking position is highly required. If the passenger flow density is high, the train door may be crowded, and the temperature in the long-term running tunnel is difficult to solve. The new airport line in Hong Kong, the disembarkation lines in Shenzhen, Guangzhou Metro Line 2 and all the underground lines after it, Guangfo Metro, Shanghai Metro except Line 2, Hangzhou Metro Line 1, Suzhou Metro Line 1, Chongqing Metro Line 1, Chengdu Metro Line 1, Changsha Metro Line 1 and most of the underground lines built in China in recent years are adopted by the system.

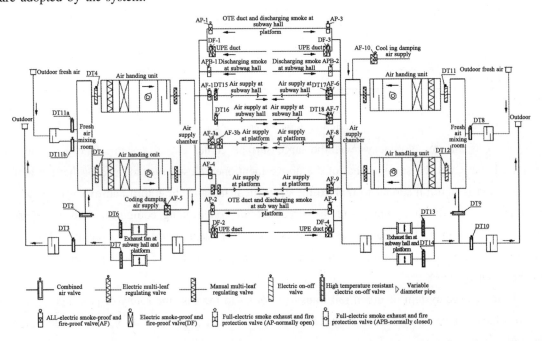

Figure 4-2 Principle of air conditioning system in station public area

In Singapore, Malaysia, Japan, France, the United Kingdom, the United States and Denmark, Applications in these countries and regions fall broadly into two categories: One type is tropical and subtropical region with hot climate. The main purpose of the screen door system is to simplify the station air conditioning and ventilation system and to save energy and reduce project investment. The screen door system is completely closed on the platform, such as Singapore NEL Line, Hong Kong New Airport Line, Tseung Kwan O Line and so on. The other is in non-hot areas, the main purpose of using screen doors is to consider the safety of passengers waiting for a bus, mainly in unmanned urban rail transit system or high-speed train stations, such as France's Toulouse light rail system, Paris Line 14 for unmanned system.

4.2.2 Effect Evaluation of Application of Each System

Advantages of the screen door system are that it creates a safety barrier that prevent passengers from unintentionally or intentionally falling onto the track, and isolates the impact of the train noise on the platform. In addition, the cooling capacity of the screen door system in the station of the same scale is about 2/5 of that in the station without the screen door system, and the area of the corresponding environment-controlled computer room can be reduced by about 1/3, so the annual operation cost is only half of that of the closed system. However, the installation of the screen door requires a large investment, and then increases the maintenance workload and costs of the screen door, and the existence of the screen door will affect the platform level roadway wall advertising effect, the platform has a sense of narrowness, especially for the side platform.

The advantage of the closed system is that the design temperature and air velocity in the station and section tunnel meet the design requirements under different working conditions. The conversion of environmental control method is simple, the platform view is wide, the advertising effect is good, but the cooling capacity is large, the area of environmental control room is large, and the energy consumption is high. In addition, the platform environment is affected by train noise.

The open-type ventilation system is mainly used in the north of China, and is not suitable for hot summer and cold winter and warm summer and warm winter areas in China. Closed system and screen door system are widely used in hot summer and cold winter, hot summer and warm winter areas. Occasionally large surface cooler closed system appears.

The advantages and disadvantages of the ventilation and air conditioning system of urban rail transit are shown in Table 4-1.

Comparison of advantages and disadvantages of urban rail transit air conditioning　　Table 4-1

System	Description	Advantages	Disadvantages	Scope of application
Open system	The action of a piston or mechanical ventilation by which an underground space is ventilate with that outside world through an air booth	Simple system, less equipment, simple control and low energy consumption	Low standard, unable to effectively control the environment inside the station and organize fume control and exhaust	Old Line in Northern Europe and America, Line 1 and Line 2 in Beijing, China

continued

System	Description	Advantages	Disadvantages	Scope of application
Close system	According to the change of outdoor climate, the tunnel ventilation facilities and the operation method of the tunnel ventilation system can be controlled by open and closed air valves. The air at that station communicate with the tunnel	The piston effect induces the air of the station into the tunnel to reduce the temperature. The air temperature in the interval tunnel is lower than that in the screen door system under the same operation conditions. Platform view is broad, advertising effect is good	The temperature field and velocity field of the station can not maintain stability, and the station air quality is difficult to control. When passengers fall into the track due to accident or special circumstances, it will have a serious impact on the normal operation; High investment and operation cost of air-conditioning system in air-conditioning season; Large ventilation and air conditioning system room; Large investment in civil engineering	Cities north of the Yangtze River in China
Screen door system	Based on the closed system, screen doors are used to isolate the station from the tunnel area	Improving safety; Reduce the influence of piston effect on station, reduce the air convection between station and tunnel, reduce the loss of station cooling load, improve the station air cleanliness, and reduce the noise brought by train approaching station; Savings on initial investment, operation costs and initial investment in civil engineering of ventilation and air conditioning systems	Increased initial investment and operation costs; Increase the Cooperation relationship with related specialties; The piston effect drains the hot air from the interval tunnel to the outside and introduces fresh air into the outdoor cooling tunnel. It is difficult to control the temperature in the tunnel during the hot season	Yangtze River Basin and Cities to the South in China

According to *Code for design of Metro* (*GB 50157—2013*) issued by China in 2013, "The ventilation and air conditioning system of metro should ensure that the air quality, temperature, humidity, air distribution, air velocity and noise of the internal air environment can meet the requirements of the physiological and psychological conditions of the personnel and the needs of the normal operation of the equipment."

In the design of urban rail transit, in determining the outdoor calculated dry-bulb temperature of the summer air-conditioned fresh air, The dry-bulb temperature of 30 hours per year is not guaranteed during the late peak load of metro in recent 20 years, but the average temperature of 50 hours per year is not guaranteed according to *Code for Design of Hating Ventilation and Air Conditioning* (GB 50019—2003) (hereinafter referred to as *HVAC Code*), because the *HVAC Code* is mainly aimed at ground construction projects, which is different from the situation of metro. According to the statistics of urban rail transit operation data, the passenger load of urban rail transit is relatively low, which is only 50%-70% of the late peak load. If the air conditioning load is calculated according to the standard, it can not meet the requirements of the late peak load of urban rail transit. If the air-conditioning cooling load is calculated by using the dry-bulb temperature not guaranteed for 50 hrs in summer and the late peak load of urban rail transit at the same time, the two peaks will be superimposed and the air-conditioning load will be on the high side. Therefore, it is more reasonable to adopt the outdoor temperature corresponding to the time when the metro evening peak load occurs.

The maximum average temperature of the hottest month and day in the normal working condition of the interval tunnel is $f \leqslant 35℃$. The temperature standard for train blocking conditions is $f \leqslant 40℃$. In order to keep the train air conditioner condenser running normally, fresh air must be fed into the section tunnel by the TVF (Tunnel Ventilation Fan) fan of the rear station of the train, and the air in the section tunnel is discharged to the ground by the TVF fan of the front station section tunnel, and the air flow direction in the section tunnel is consistent with the direction of the train. As the condenser heat production of the train blocked in the interval tunnel is continuously released into the ambient air, the air temperature around the train is rapidly increased due to the stop of the train piston air, and when the air inlet temperature of the train air conditioner condenser is higher than 46℃, some compressors will be unloaded; When the intake air temperature is greater than 56℃, the compressor stops rotating and the temperature and humidity in the train will be unbearable to the passengers. As that air temperature around the air condition condenser on the train top is 5-6℃ higher than the air temperature around the train, the air temperature around the train is required to be lower than 40℃ in order to make the air temperature around the condenser lower than 46℃.

The relative humidity of the station is controlled at 45%-65%. Minimum fresh air volume for personnel: urban rail transit project is underground project, and the air quality inside the station is worse than that outside, so the fresh air volume standard for personnel is particularly important. According to the regulations and taking into account the specific conditions of each place, the platform air-conditioning season of the station hall shall adopt the air volume of not less than 12.6 $m^3/(h \cdot person)$ per passenger, and the fresh air volume shall not be less than 10% of the total air volume of the system; For non-air-conditioned seasons, each passenger shall press not less than 30 $m^3/(h \cdot person)$, and the ventilation frequency shall be greater than 5 times/h; The fresh air volume of the personnel in the equipment management room shall be not less than 30 m³/(h · person) and not less than 10% of the total air volume of the system.

The air quality standard is CO_2 concentration less than 1.5 ‰. When the noise control standards are in normal operation, the common area of the station hall and platform shall not be more than 70 dB (A); Ground air kiosk ≤ 70 dB (A) in daytime and ≤ 55 dB (A) at night; Environment control computer room ≤ 90 dB (A); Management room (studio and lounge) ≤ 60 dB (A). In terms of air distribution in the common area of the station hall and platform layer. As the urban rail transit station is a rectangular limited space, it has a large calorific value. It is required to supply air uniformly along the length of the station, the air return port should also be arranged at the upper part, therefore, the typical island-type station adopts the form of two-supply-one-return or two-supply-two-return with two-supply-one-return from the top to the bottom on both sides of the station hall and above the platform. The supply duct is arranged on both sides of the station hall and above the platform, and the air outlet is uniformly distributed downward. The air return pipe is arranged at the middle and upper part of the station, as shown in Figure 4-1. Side platform is a one-free-one-return form. The platform exhaust is composed of the train top exhaust and the platform bottom exhaust: the train top exhaust duct is arranged above the train track, and the train top exhaust port corresponds to the position of the train air conditioner condenser; the supply and exhaust duct under the platform is a civil air duct, and the air outlet under the platform corresponds to the heating position under the train. The exhaust duct on the top of the train is also used as the exhaust duct.

The wind speed design standard shall be set according to the normal operation conditions and the accident ventilation and fume exhaust conditions.

Under normal operation conditions, the wind speed of the structural air duct and air shaft shall not be greater than 6 m/s; The wind speed of the air outlet is 2-3 m/s; The wind speed of the main air duct shall not be greater than 10 m/s; The wind speed of branch duct without supply and return port is 5-7 m/s, and the wind speed with supply and return port is 3-5 m/s; The wind speed of the grille of the wind kiosk shall not exceed 4 m/s; Wind speed between muffler plates is less than 10 m/s.

Under the condition of accident ventilation and fume exhaust, the wind speed of the tunnel is controlled at 2-11 m/s; the wind speed of the main exhaust duct is less than 20 m/s (metal pipe is adopted); the wind speed of the main exhaust duct is less than 15 m/s (non-metal pipe is adopted); the wind speed at the exhaust port is less than 10 m/s. The main design standards for disaster prevention include: only one fire in urban rail transit is considered; fume discharge from the fire in the station hall shall be calculated according to 1 m³/(min · m²); fume discharge shall be calculated according to the downflow velocity at the staircase passage from the station hall to the platform at least 1.5 m/s for platform fire; fume discharge is calculated by wind speed of 2-2.5 m/s in the cross-section of a single tunnel.

4.3 Internal Ventilation System of Metro Station

The composition of ventilation and air-conditioning system in urban rail transit is closely related to the division of the functional areas of the lower stations, in which safety considerations must be taken into account, such as the installation of fume control and exhaust systems. The ventilation and air-conditioning system inside a station can be simplified into four subsystems, whether the platform is equipped with a screen door system or a closed system as is commonly known:

(1) Ventilation, air conditioning and fume exhaust system in public areas;
(2) Ventilation, air conditioning and fume exhaust system for equipment management room;
(3) Tunnel ventilation and fume exhaust system;
(4) Air conditioning refrigeration water circulating system.

4.3.1 Ventilation and Air Conditioning in Public Areas

The public area of the station hall and platform layer of the urban rail transit station is the main place for passengers' activities, and also the main control area for the air conditioning and ventilation of the environmental control system. Ventilation and air-conditioning systems in public areas are simply referred to as large systems. In the design, ventilation ducts shall be arranged within the length range of the station hall and platform to uniformly supply and exhaust air, A roof return and exhaust duct (OTE) is arrange on that top of the train on the platform layer, a platform down net and an exhaust duct (UPE) are arrange on the lower part of the platform layer, and a centralized supply port is arranged on the station end of the train inbound end, the purpose of which is to cool the hot air in the inbound station as soon as possible, increase air disturbance and reduce the influence of the piston air on the passengers. The principle of the large air conditioning system in the common area of the station is shown in Figure 4-2.

The air conditioners and ventilators of the station are arranged on the floor of the station hall at both ends of the station, equipment is arranged symmetrically, basically half of the station's load is borne by each station, The main systems of the station are: four combined air conditioning units, four return and exhaust fans, the utility model has the function of removing the residual heat and moisture in the common area of the station through air conditioning or mechanical ventilation, To create a comfortable riding environment for passengers, and in case of fire, fume is exhausted by mechanical ventilation, so that negative pressure area is formed in the station. Fresh air enters the station hall and platform from outside through pedestrian passageway or staircase entrance, which is convenient for passengers to evacuate and firefighters to put out fires.

The air conditioner of the station hall layer adopts the form of up-sending and up-returning, and the platform layer adopts the form of up-sending and up-returning combined with down-returning. Generally, the rail top return is arranged on the top of the train, and the exhaust air duct takes away the heat dissipation of the train air conditioner condenser directly from the return air. At the same time, the lower return and exhaust ducts of the platform are arranged under the platform, so as to directly take away the heat and dust of the electric appliances and brakes under the train with the return air.

In case of fire on the station platform or train, the equipment of other station large systems will stop running except the platform return and exhaust fan of the station, so that a downward air flow of not less than 1.5 m/s will be formed between the upper and lower passages of the station platform and the station hall, which is convenient for passengers to withdraw to the station hall and the ground in the face of the air flow; In case of fire in the station hall, all the return and exhaust fans of the station hall start to exhaust fume, and other equipment of the large system stops running, so that the exit and entrance passages form downward air flow from the ground to the station, which is convenient for passengers to evacuate to the ground against the air flow.

4.3.2 Ventilation and Air Conditioning of Equipment Management Room

The management and equipment rooms of the station are mainly distributed with various operation and management rooms and equipment rooms of the control system. Its working environment will directly affect the safe and normal operation of urban rail transit, in fact, it is the core area of urban rail transit station management system, but also the key area of environmental control system design. This kind of housing according to the different needs of each station and set. Ventilation and air-conditioning system of station equipment room is also referred to as small system. The computer room is generally arranged in the station hall, platform layer at both ends of the station. The station hall layer mainly concentrates on connecting, signal, environmental control electronic control room, low-voltage power supply, environmental control computer room and station management room. The platform layer mainly arranges high-voltage power supply room and medium-voltage power supply room. The principle of ventilation and air conditioning system for station equipment management room is shown in Figure 4-3.

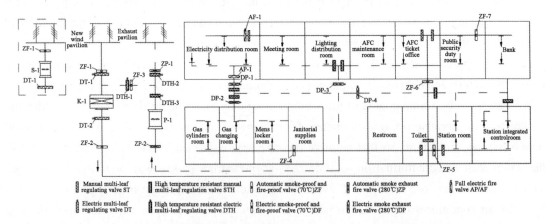

Figure 4-3 Principle of ventilation and air conditioning system in station equipment management Room

Due to the different equipment environment requirements and temperature and humidity requirements of various rooms, independent supply and (or) exhaust systems are basically set up for the air conditioning and ventilation of the small system according to the following four forms according to the different requirements of various rooms:

(1) Houses to be air-conditioned and ventilated, such as connecting, signal, station control, environment-controlled electronic control, meeting rooms, etc.

(2) rooms that only need ventilation, such as high and low voltage rooms, lighting and power distribution rooms, environmental control rooms, etc.

(3) rooms that only require ventilation, such as toilets, storage rooms, etc.

(4) The rooms to be protected by gas extinguishing, such as connecting, signal equipment room, environmental control electronic control room, high and low voltage room, etc.

The equipment composition of the station small system mainly includes axial flow fans serving station equipment and management rooms, Cabinet and hanging air-conditioning units and various air valves are used to provide a comfortable working environment for management and staff and a normal operation environment for all kinds of equipment by controlling the temperature and humidity of each room and other environmental conditions. When a fire occurs, fume is exhausted by mechanical ventilation, which is beneficial to the evacuation of workers and the extinguishment of firefighters. Close the supply and exhaust ducts in the gas extinguishing room to extinguish the fire in a sealed manner.

4.3.3 Tunnel Ventilation and Fume Exhaust System

The equipment of the tunnel ventilation system is mainly composed of four tunnel ventilators respectively arranged in the station halls at both ends of the station and the platform, and their corresponding components such as mufflers, combined air valves, air ducts, air shafts and air kiosks. Its function is to eliminate the residual heat and moisture in the interval tunnel by mechanical supply, exhaust or train piston wind, and ensure the normal operation of the train and the equipment in the tunnel. A typical section ventilation and fume exhaust system is shown in Figure 4-4. In addition, the tunnel fan is turned on half an hour before the operation every morning for cooling and ventilation, which can not only make use of the fresh cold air in the morning to exchange air and cool the urban rail transit, but also inspect and maintain the equipment in time to ensure that it can be put into use in case of accidents. When the train stays in the tunnel for various reasons, and the passengers do not get off the train, the train runs along the direction of mechanical ventilation, cooling the train air conditioning condenser and so on, so that the passengers still have a comfortable travel environment. In case of train fire, every effort shall be made to make the train run within the platform of the station so as to facilitate the evacuation of personnel and extinguish fume. When the train in the fire was unable to reach the station and was supply to stop in the tunnel, The tunnel fan at one end of the tunnel shall deliver fresh air to the fire site, and the tunnel fan at the other end shall exhaust fume from the tunnel to guide the passengers to evacuate the accident site in the direction of air flow, and the fire fighters shall carry out fire extinguishing and rescue work in the direction of air flow.

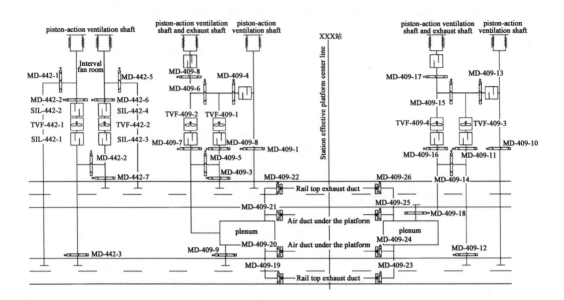

Figure 4-4 Principles of typical interval systems

In addition, the tunnel ventilation system also includes the equipment at the tunnel entrance of the closed system and local ventilation facilities at the turnaround line of the transition section. Tunnel entrance and station entrance are the places where the outside atmosphere communicates directly with the underground space of urban rail transit, In order to reduce the influence of high temperature air on the air conditioning system of urban rail transit, an air curtain isolation system is set up from the ground to the tunnel entrance. The system is composed of two wind turbines and air curtain nozzles. The machine room is set up at the tunnel entrance. Turnouts are arranged at both ends of the turnout line to connect with the main line. The turnout line is generally in the middle of the main line with a large cross-sectional area. It is difficult for the tunnel ventilators in the original station to satisfy the ventilation of the main line and the turnout line at the same time, and the additional fan will increase the area of the machine room, which is also difficult to implement. Through comparison of various schemes, jet fan ventilation scheme is often adopted, which is jointly organized by jet fan and station tunnel fan. The main purpose of this design is to solve the problem of underground space tension and the difficulty of air distribution on the turnaround line (transition section).

4.3.4 Air Conditioning Refrigeration Water Circulating System

The function of station air-conditioning refrigeration water circulating system is to make cooling source for station air-conditioning system and supply it to air treatment equipment (combined narrow box, cabinet fan coil) in station air-conditioning large and small systems, and to send heat out of station through cooling water system.

At present, the ventilation and air conditioning system of urban rail transit is divided into independent cooling and centralized cooling according to the configuration relationship between the cooling source and the station.

1. Independent cooling

Generally, each underground station is equipped with an independent refrigeration station, which is usually operated by a combination of two screw chillers with the same refrigeration capacity (refrigeration capacity ≥ 1000 kW) and one screw chiller with the same refrigeration capacity (refrigeration capacity ≤ 500kW) (or piston chillers and other forms). Two screw units with large refrigerating capacity are selected according to the cooling load of large system air conditioner. A screw chiller with small refrigerating capacity is selected according to the cooling load of the air conditioner of the small system (responsible for equipment management room). It can be operated either alone or in combination with the large system and run in combination with the large capacity screw chiller. Air conditioning water system also includes refrigeration, cooling water pumps, cooling towers, air conditioning boxes and other end equipment. The principle of air conditioning water system is shown in Figure 4-5.

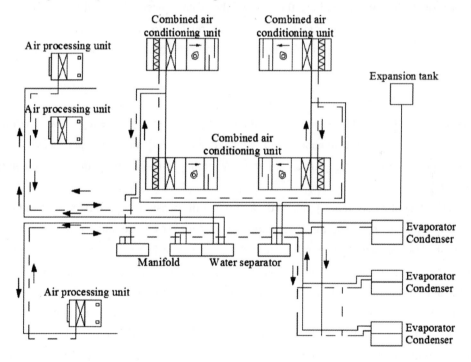

Figure 4-5 Principle of air conditioning water system

In the drawing, the number of chilled water pumps and cooling water pumps corresponds to the number of chillers one by one. The diverter of small system is connected with the diverter of cold source in the common area through ducts. Valves are arranged on the connecting ducts, which are closed during normal operation and manually opened when they need to be reserved for each other. The refrigeration station is centrally arranged in the refrigerator room at one end of the station, as close to the load center as possible, and tries to shorten the length of the refrigerated water supply/return pipe.

Air conditioning chilled water temperature: 7℃ for water supply and 12℃ for backwater. Cooling water temperature: 32℃ for water supply and 37℃ for backwater. The refrigerated water system adopts the primary pump system. The return pipe of the small system air conditioning unit is provided with an electric two-way valve. The pressure difference bypass valve is arranged between the small system water collector and the water distributor. The large system water collector and the water distributor are not connected.

The constant pressure of chilled water system adopts expansion water tank.

Under the normal operation condition of air conditioning season, the number of large capacity screw chillers and small capacity screw chillers is controlled according to the cooling load of the station. In non-air-conditioned seasons, all water systems are out of service. The water system shall continue to operate according to the normal operation conditions at that time when the block accident of the interval tunnel occurs. When a fire occurs in the common area of the station hall and platform or in the tunnel section, the part of the water system that is the cold source of the large-scale system shall be closed and only the part related to the small-scale system shall be operated. When a fire breaks out in the equipment room of the small system, the water system stops running completely.

2. Centralized cooling

Centralized cooling system has the advantages of high energy efficiency, low environmental thermal pollution, easy maintenance and management. As an important way of energy conservation and environmental protection, it is becoming a major trend in urban planning and development.

Centralized cooling system is adopted in urban rail transit lines. Firstly, the influence of cooling towers on the surrounding environment is reduced by rationally arranging the cooling stations in the rail network. Secondly, it reduces the coordination workload with the urban planning department in order to meet the requirements of outdoor cooling tower equipment occupation and aesthetics. Thirdly, the number of refrigeration stations is reduced and the underground limited space is saved. Fourthly, the operation efficiency is improved, and the centralized maintenance and management is facilitated, and the automation level is improved. The centralized cooling system has been successfully applied in Guangzhou Metro Line 2, Hong Kong Metro Station, and Cairo Metro Station.

The centralized cooling system of urban rail transit shall be equipped with centralized chillers, linkage equipment and other auxiliary equipment, and cold water pipes shall be laid through outdoor pipe corridors, underground trenches and section tunnels, and the cold water shall be conveyed to the end of the station air conditioning system by secondary water pumps. The principle and process of centralized cooling system are shown in Figure 4-6.

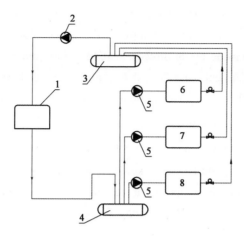

Figure 4-6 Principle of the cooling system

1-cooling tower;2-primary pump for cool water;3-water collector;4-water separator;5-secondary pump for cool water;
6-Station 1 refrigeration chiller;7-Station 2 refrigeration chiller;8-Station 3 refrigeration chillers

4.4 Operation Status of Metro Ventilation and Air Conditioning System

The operation of ventilation and air conditioning system in urban rail transit can be divided into two states: normal operation and congestion and fire accident operation. Corresponding to these two states, the system can be divided into normal operation method, congestion and fire accident operation method.

4.4.1 Normal Operation

1. The air conditioning and ventilation system of the station

Under the new air conditioning and ventilation operation environment, the enthalpy value of the outside atmosphere is smaller than the station air enthalpy value. Start the refrigeration air conditioning system and run the brand-new fan. The outside air is cooled by the air conditioner and sent to the station hall and the common area of the platform, and the exhaust air is discharged from the ground completely. This operation method is called fresh air conditioning and ventilation operation.

Under the condition of small new air conditioning and ventilation, start up the refrigeration air conditioning system, run the new air conditioning fan, part of the return air/exhaust air is discharged to the ground, part of the return air is mixed with the external fresh air delivered by the new air conditioning fan, and sent to the common area of the station hall and platform after being cooled by the air conditioner. This operation method is called small fresh air conditioning and ventilation operation.

Under the non-air-conditioned ventilation operation environment, the enthalpy of the outside atmosphere is less than or equal to the enthalpy of the air-conditioned supply, the refrigeration system is shut down, the outside air is directly sent to the common area of the station hall and platform without cooling treatment, and the exhaust air is completely discharged from the ground. This operation method is called non-air-conditioned ventilation operation.

2. Ventilation system of section tunnel

In the natural closed system, $i_{out} \geq i_{in}$, close the tunnel ventilation shaft, open the circuitous air passage in the station, and ventilate and exchange air in the section tunnel by the action of the piston of the train. The piston air enters the tunnel from the station behind the train, and the air flow part in front of the train enters the station. Partially circulates from the circuitous duct to parallel adjacent tunnel openings.

In the natural open system, $i_{out} < i_{in}$, open the tunnel air shaft; By the action of the train piston, the outside air enters the urban rail transit tunnel from the tunnel ventilation shaft behind the train operation. By the action of the train piston, the outside atmosphere is drained from the ventilation shaft of the tunnel in front of the train.

In the mechanical open system, $i_{out} < i_{in}$, the natural open system can not meet the requirements of temperature and humidity in the tunnel, the tunnel ventilator starts and carries out mechanical ventilation; the outside air is sent from the tunnel ventilation shaft behind the train operation to the tunnel through the tunnel ventilation fan, which is the supply method. Outside air is discharged from the tunnel ventilation shaft in front of the train to the ground through the tunnel ventilation fan, which is an exhaust method.

In summary, it can be seen that the operation method and ventilation method of the tunnel ventilation system is a more complex problem, it is not completely independent of the system, and has a lot of connections with the station system, the operation of the tunnel ventilation system and the station system will act together.

4.4.2 Operation of Obstruction and Fire Accidents

1. Blocking accident operation

Congestion accident operation refers to the train in normal operation because of various reasons to stay in the tunnel, at this time passengers do not get off the train, this condition is called congestion accident operation.

In the station air conditioning and ventilation system, the station air conditioning and ventilation system shall operate normally when the train is blocked in the section tunnel, and the station air conditioning and ventilation system shall operate with fresh air when the TVF fan is required to operate. When running the TVF fan, the platform returns and the exhaust fan stops running, so that the cold air of the station is sent to the tunnel blocked by the train through the TVF fan.

In the ventilation system of the section tunnel, when the station runs naturally in the closed mechanical environment, if the train blocks the operation of TVF fan in the section tunnel, the station cold air is sent to the tunnel. In the open mechanical operation environment, when the station starts to run, if the train is blocked in the tunnel, the TVF fan operates in the mechanically open method.

2. Fire accident operation

The space of metro is narrow. Once a fire occurs, the passenger evacuation and fire-fighting conditions are worse than those on the ground. Therefore, the design should be taken as the key problem to be solved. In case of fire, all operation and management shall be strictly subject to the needs of passenger evacuation and rescue work. Fire accidents include tunnel fire and station fire, and station fire includes train fire, platform fire and station hall fire.

When a train is caught in a fire in a tunnel, First of all, consider driving the train into the station. If the train is parked in the section, determine the fire position and stop position of the train, deliver fresh air and exhaust fume to the fire site according to the fire operation method, let the passengers evacuate the accident site in the direction of fresh wind, and let the firefighters enter the scene to put out the fire.

In case of train fire and plat form fire, a downward air flow of not less than 1.5 m/s shall be formed between the upper and lower passages from the platform to the station hall, and the passengers shall be evacuated from the platform to the station hall and the ground against the air flow. Therefore, the equipment of other station large systems shall be stopped except that the platform return and exhaust fans of the station are operated to exhaust fume to the ground. In case of fire in the station hall, the return and exhaust fans of the station hall are all started to exhaust fume, and other equipment of the large system is stopped, so that a downward air flow from the ground to the station is formed at the entrance and exit passage, and the passengers are evacuated to the ground in the direction of the air flow.

4.5 Calculation of Metro Ventilation and Air Conditioning System Load

4.5.1 Calculation of Screen Door System Load

In a screen door system, screen doors separate the tunnel from the station platform, The air conditioning load of the station is relatively less affected by the tunnel, and the heat dissipation in the common area of the station no longer includes the heat generated by train driving equipment, train air conditioning equipment and mechanical equipment, but only the heat dissipation of the station personnel, lighting and equipment, the heat transfer of the temperature difference between the platform and the platform, and the heat brought by the permeated air. Compared with the closed system, it has less influence of train and tunnel piston wind on the station, greatly reduces the cold load and the complexity of the system, and the load calculation is relatively simple.

1. Human body heat load

Station personnel are divided into fixed personnel (including station staff, commercial service personnel, etc.) and mobile personnel (mainly urban rail transit passengers). The number of fixed personnel is basically stable hourly throughout the day, and the calorific value is calculated according to the metabolic rate of the human body in the state of sitting (or standing) sales, and the average residence time is calculated according to the working time. The number of mobile personnel changes hourly throughout the day, the number of peak hours is large, calorific value calculation reference walking (or standing) state of the human body metabolic rate.

Therefore, the determination of human body heat load is the key to the determination of passenger flow, which is generally derived from the passenger flow forecast report of the local traffic planning department, and the peak hour passenger flow of the station area should be considered in the calculation. According to the information and some data, the stop time of boarding passenger flow at the station is 4 minutes, in which the passengers enter the urban rail transit station hall from the ground to stay for 1.5 minutes, and the platform waiting time is 2.5 minutes. The stop time of alighting passenger flow station is about 3 minutes, and the average time of this process is related to the train running interval. When the stay time of boarding and alighting passengers in the station is determined, considering the appropriate clustering coefficient, the heat dissipation load of the station is determined.

2. Electrical and mechanical load

The heat dissipation of lighting equipment, advertising light boxes, escalators, vertical elevators, guide signs and ticket sales (inspection) machines can be calculated conveniently from the actual power of various electrical facilities.

3. Heat transfer load of screen door

The screen door isolates two different temperature environments, and the heat transfer between the station environment and the tunnel can be calculated by one-dimensional steady-state heat conduction. After determining the area and material of the station screen door, the heat transfer load of the screen door is determined.

4. Infiltrate the heat from the wind

This part of the maximum heat, the station's total cooling load is also the greatest impact. According to the past design experience, the permeable air at the station entrance and exit is calculated as 200 W/m^2 (cross-sectional area), and the air leakage at each station of the screen door is estimated as 5-10 m^3/s.

5. Wet load

It can be divided into the moisture dissipation of personnel, the moisture dissipation of the structure wall and the moisture dissipation brought by the permeable wind. According to the empirical calculation of relevant data, the moisture dissipation of the station side wall, roof and floor is 1-2 g/(m² · h); The moisture dissipation of personnel was 193 g/h at 27 ℃ during light work.

4.5.2　Calculation of Closed System Load

When the screen door is not installed in the station hall, the factors influencing the energy consumption system of the station air conditioning system are complicated. In addition to the parameters listed above, the influence of vehicle running (such as departure density, number of running queues, parking time, traction curve, etc.) should be considered. At this time, the load caused by train running heat dissipation becomes the main source of the station air conditioning load. In addition, because there is no screen door, it is difficult to treat station and tunnel differently in air conditioning load calculation.

There are many methods to calculate the air conditioning load of closed system, but at present they only stay at the level of estimation, and the accuracy of each method is quite different. The following is a simple estimation method cited in the paper " Discussion on Environmental Controlled Ventilation of Metro" for reference.

1. Train heat production

Train heat generation is the main component of urban rail transit waste heat.

If Q_1 is the train thermal output (kW), then

$$Q_1 = 2N_0 n_g n_i (G_i + g_p n_p) L \qquad (4-1)$$

Where: L——The length of the calculated section of train travel(km);

g_p——Average weight per person (t/person);

n_p——The calculated number of persons per vehicle (person/section);

G_i——Weight of each vehicle (t/section);

n_i——The formation of each train (sections/trains);

n_g——Train running density (logarithm of trains per hour) (pairs/h);

N_0——1 t · km of train power consumption [(kWh/(t · km)].

In calculating heat production, 70% of the maximum density is preferable, which is generally larger than the average value and is generally calculated according to the average power consumption per ton-kilometre of operation [0.05-0.07 kWh/(t · h) in Japan and 0.052 kWh/(t · km) in the Soviet Union]. If there is air-conditioning equipment on the train, in addition to the above heat generation, there should be additional air-conditioning equipment heat generation.

2. Lighting heat production

Electric lighting heat generation Q_2, which is calculated as follows:

$$Q_2 = N_a A + N_1 l \qquad (4-2)$$

Where: N_a——Lighting load per unit area of the platform of the station hall (kW/m²);

A——Platform area of the station hall (m^2);

N_1——Lighting load per metre of interval tunnel (kW/m);

l——Length of the interval tunnel section (m).

If a fluorescent lamp is used, the lighting charge shall also include the amount of power consumed by the ballast.

3. Personnel heat production

The heat output of personnel is Q_3, which includes the personnel on the production station and the personnel on the train.

$$Q_3 = q_p (\Sigma b + 2n_g n_j n_p) \frac{L}{v} \tag{4-3}$$

Where: v——Train speed (km/h);

L——Calculated section length of section tunnel (km);

Σb——Calculate half of the total number of persons at the two adjacent stations in the interval (persons);

q_p——Human calorie production (kW/person).

The heat produced by human body is composed of sensible heat and latent heat, and the residual heat is calculated according to the total heat.

When the train is air-conditioned, the heat generated by the condenser replaces the heat generated by the people on the train, which is generally 1.5 times of the heat generated by the people on the train.

4. Heat production of power equipment

The heat output of the power equipment is Q_4, and the calculation formula is:

$$Q_4 = N_w \tag{4-4}$$

Where: N_w——The kilowatts of heat-emitting power equipment, including electrical machines and other power equipment in urban rail transit systems (kW).

Q_4 should pay attention to the following problems: in the ventilation system, only the heat generated by the motor of the supply equipment is considered, but the heat generated by the motor of the air exhaust equipment is not included; The heat dissipation amount of the drain pump is not included in the calculation because it is removed by water; The heat generated by the equipment in the production room and the equipment room shall be considered by the local ventilation system and shall not be included in the calculation.

5. Heat absorption of the tunnel wall

The heat absorption and release of the tunnel wall in urban rail transit system depends on the temperature of the ground around the tunnel. When the air temperature in the urban rail transit system is higher than the surface temperature of the tunnel wall, the tunnel wall absorbs heat. When the air temperature in the urban rail transit system is lower than that in the tunnel wall, the tunnel wall exotherms heat. These calories are Q_5.

$$Q_5 = KF\Delta t \tag{4-5}$$

Where: K——Heat transfer coefficient $[kW/(m^2 \cdot \text{℃})]$;

F——The contact area between the lining structure and the surrounding strata (m^2);

Δt——Difference between the mean temperature t_1 of the tunnel and the calculated The thermal conductivity K is related to a number of factors, such as lining material and thickness, the nature of surrounding strata, the state of groundwater, etc., which can generally be determined by the following formula:

$$K = \frac{1}{\dfrac{1}{\alpha} + \dfrac{l_c}{\lambda_c} + \dfrac{l_e}{\lambda_e}} \tag{4-6}$$

Where: α——Convective heat transfer coefficient of wall air to tunnel lining surface $[(kW/(m^2 \cdot \text{℃})]$;

λ_c, λ_e——The thermal conductivity of the lining and the surrounding formation, the value of which is related to the properties of the material $[(kW/(m2 \cdot \text{℃})]$;

l_c——Average thickness of concrete lining (m);

l_e——The thickness of the medium in the part where the ambient temperature changes (m).

l_e is the distance from the surface of the lining to the point where the temperature in the soil no longer varies. Because urban rail transit is an underground structure, the temperature of the surrounding stratum has not changed dramatically. During the initial period of operation, the heat released from the tunnel is transferred to the stratum, and the phenomenon of heat dissipation occurs in the stratum. After a certain period of time, the temperature is fixed at a certain distance from the ground on the inner surface of the tunnel. This distance is related to groundwater and soil quality, and is generally considered as 0.5 m in approximate calculation. The calculated temperature of the surrounding strata is calculated according to the annual average temperature of the strata, and the groundwater temperature is generally adopted for the water-bearing strata.

The above-described heat generation and wall heat absorption and release in urban rail transit are various, so the residual heat Q in urban rail transit system is:

$$Q = Q_1 + Q_2 + Q_3 + Q_4 - Q_5 \tag{4-7}$$

Because the heat sources of different locations in urban rail transit system are different, and with the different operation years, even the heat generation at the same location also changes. Therefore, the detailed calculation needs a computer program to simulate the calculation.

Exercise

4.1 What requirements must the environmental control system of urban rail transit lines meet?

4.2 What are the ventilation and air conditioning systems of urban rail transit, and what are the advantages and disadvantages of each system?

4.3 What are the components of the ventilation and air conditioning system of the metro station?

4.4 What is fresh air air conditioning, ventilation operation?

4.5 Explain the operation principle of tunnel ventilation system in urban rail transit.

 Key vocabulary:

air curtain　空气幕
closed system　闭式系统
fume exhaust system　排烟系统
minimum fresh air requirement　最小新风量
piston air shaft　活塞风井
side platform　侧式站台
mini-fresh air conditioning　小新风空调

non-air-conditioning ventilation　无空调通风
all-fresh air conditioning　全新风空调
environmental control system　环控系统
heat and humidity exchange　热湿交换
open system　开放系统
screen door system　屏蔽门系统

Chapter 5 Operation Ventilation for Other Underground Space

[Important and Difficult Contents of this Chapter]
(1) Air environmental standards and air-conditioning and ventilation systems for underground shopping malls.
(2) Pollutant composition, air standard and ventilation method of underground parking lot.
(3) Standard of air composition and air concentration in the mine and ventilation method in the excavation of the mine.
(4) Air environment standard, ventilation method in peacetime and ventilation method in wartime of civil air defense project.

5.1 Operation Ventilation of Underground Shopping Malls

In recent years, the economic development of the city has led to the rise of land prices and the shortage of land on the ground, the construction cost of the above-ground shopping malls is relatively large, so the construction of underground shopping malls began to rise. A good air environment is one of the preconditions to ensure the effectiveness of underground shopping malls, so the ventilation design of underground shopping malls is very important.

5.1.1 Selection of Design Parameters for Underground Shopping Malls

1. Temperature and humidity standard

Temperature and humidity are important indexes of the air environment in underground space. Table 5-1 lists the temperature and humidity parameters specified in the domestic and international standards.

Temperature and humidity parameters　　　　Table 5-1

Standards	Summer		Winter	
	t_w (℃)	j_w (%)	t_d (℃)	j_d (%)
Code for design of civil air defense works (*GB 50225—2005*)	≤30	≤70	≥16	≥30

continued

Standards	Summer		Winter	
	t_w(℃)	j_w(%)	t_d(℃)	j_d(%)
Code for design of civil air defense basement (GB 50038—2005)	≤28	≤75	≥16	≥30
Code for design of store buildings (JGJ 48—2014)	26~28 Artificial cold source 28~30 Natural cold source	55~65 Artificial cold source 60~65 Natural cold source	16~18	30~35
Japanese underground street	24~26	50~60	18~22	50

2. Minimum fresh air volume standard

Refer to Table 5-2 for minimum fresh air volume standard of air conditioning in underground shopping malls.

Minimum fresh air volume standard　　　　　　　Table 5-2

Standards	The minimum fresh air volume standard ($m^3/p \cdot h$)	CO_2 concentration (%)
Code for design of civil air defense works (GB 50225—2005)	15	
Code for design of civil air defence basement (GB 50038—2005)	≥15	
Code for design of store buildings (JGJ 48—2014)	8.5	0.2
Japanese underground street	30 $m^3/(m^2 \cdot h)$ Only ventilated, no air conditioning 10 $m^3/(m^2 \cdot h)$ (Air-conditioned)	

The minimum fresh air volume standards in Table 5-2 vary widely, some of which are lower for energy conservation. In order to meet the standard of CO_2 concentration, some standards of fresh air volume have been improved. There are two effective ways to improve the air sanitary environment of underground shopping malls: one is to ensure enough fresh air to control the concentration of CO_2, and then to improve the air cleanliness. The second is to add air filter to reduce the dust content and the number of floating bacteria in the air.

5.1.2　Air Conditioning System of Underground Shopping Malls

In order to ensure that the air in underground shopping malls is fresh, the temperature and humidity are appropriate, and to provide customers with a good shopping environment, most underground shopping malls are equipped with air-conditioning system.

Chapter 5 Operation Ventilation for Other Underground Space

1. Heat and humidity load characteristics of underground shopping malls

The air conditioning load of underground shopping malls is composed of human body load, lighting load, fresh air load, building load and equipment load. The largest proportion of which is generally the human body, lighting and fresh air three loads. Therefore, in order to effectively regulate the air quality of underground shopping malls, it is necessary to grasp the working intensity of air-conditioning and ventilation equipment, which depends on the size of underground shopping malls and the density of personnel. According to the energy-saving standard of public buildings, the average occupancy area of general stores is 3 m^2/person, and that of high-grade stores is 4 m^2/person. For the specific underground shopping malls, due to operation commodities, urban areas, purchasing power and other factors, the density of personnel is very different. Therefore, the determination of personnel density should be scientifically analyzed and calculated in the feasibility study of the proposed project according to the sufficient investigation statistical data and development trend.

Personnel density can be calculated according to Formula (5-1):

$$m = \frac{A \times W \times t}{T \times F} \tag{5-1}$$

Where: m——The density of personnel in the business hall (person/m^2);
    A——Coefficient 0.5-0.7;
    W——Peak passenger flow (person/day), calculated according to the measured value of local shopping malls of the same size;
    t——Customer stay in the mall (h) in shopping malls, 0.6-0.9 h in large shopping malls and 0.4-0.7 h in small shopping malls;
    T——Daily business hours (h);
    F——Business hall area (m^2).

Personnel density is the main basis of humid heat load, dust production, bacteria production, odor occurrence, related to the size of the air conditioning system and equipment capacity. Therefore, the determination of this parameter should be careful and accurate.

The air conditioning load of underground shopping malls is characterized by low heat and humidity ratio, generally around 4000 kJ/kg (995 cal/kg). There is also a small heat and humidity, air treatment is generally used to reduce the humidity after the cold and then re-heating.

2. Air conditioning system

Centralized all-air fixed air volume air conditioning system is the most widely used scheme in ventilation and air conditioning design of underground shopping malls. Usually, the primary air return system is adopted, and the design scheme is shown in Figure 5-1.

As can be seen from Figure 5-1, the air treatment conditions in summer and winter are different:

Summer working conditions: Outdoor fresh air is mixed with return air through rough filtration, intermediate effect, filtration, surface cooler and secondary heating, and the treated air is sent into the shopping mall along the heat-humidity ratio line by the blower through muffler, duct and air outlet to eliminate internal residual heat and moisture, thus reaching the design standard.

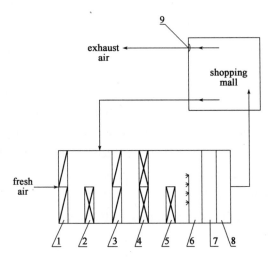

Figure 5-1   Schematic diagram of primary air return system

1-Coarse filter;2-Primary heater;3-Intermediate filter;4-Surface cooler;5-Secondary heater;6-Humidifier;7-Blower;
8-Supply muffle;9-Fume exhaust machine

Winter working conditions: outdoor fresh air is coarsely filtered. After the mixture of preheating and return air is filtered, reheated and humidified, the treated air is sent into the shopping mall along the heat-humidity ratio line by the blower through muffler, duct and air outlet, reaching the design standard.

As that design scheme of the primary air return type system has the centralize setting of the air condition room, the air can be processed under various working conditions. Easy operation and maintenance, easy control of vibration and noise; large supply, full ventilation, exhaust fan, filter season fresh air cooling and other advantages. So, it is widely used in underground shopping malls air conditioning design.

5.1.3   Ventilation System of Underground Shopping Malls

Because of the nature of the use of shopping malls, a large number of fresh air is needed, and the entrance of underground shopping malls is only an evacuation stairwell, natural ventilation will obviously make the indoor pressure too high, the need for additional ventilation system. Usually used exhaust system and fire fume exhaust system can share a set of systems. In order to ensure that fume can be exhausted effectively during fire, supply system should also be set up.

Underground shopping malls generally have lower storey heights (net heights 3.0-3.2m), and the air distribution should be carried out on the upper side and in the way of top-conveying and top-discharging. As that supply port of the upper side supply is often blocked by the shelf, the top supply is optimal. In general, the flat-feed diffuser shall be used. If the ceiling is high, the lower-feed diffuser or louver outlet shall be used. The vent is arranged at the top of the passageway or near the side wall (shelf).

5.1.4 Fume Exhaust System of Underground Mall

The fume exhaust system of underground shopping malls is set up to control the fume diffusion during the fire. Underground shopping malls use natural fume exhaust effect is poor, generally need to use mechanical fume exhaust. The mechanical fume exhaust system shall be set up separately or in combination with the exhaust system normally used. When combined, measures shall be taken to automatically convert the exhaust system to the fume exhaust system in case of fire.

5.2 Operation Ventilation of Underground Parking Lot

In recent years, with the continuous development of urban modernization, the number of cars in the city is growing rapidly, the contradiction between car storage and urban land is increasing, and the construction of underground parking lots is rising. Ventilation of underground parking lot is one of the important preconditions to ensure the normal operation of parking lot.

5.2.1 Pollutants in Underground Parking Lot and their Hazards

Pollutants emitted by vehicles in underground parking lots are mainly harmful substances such as carbon monoxide (CO), hydrocarbons (HC), nitrogen oxides (NO_x), particulate matter (PM) and so on.

1. Carbon monoxide (CO)

Carbon monoxide is produced by the imbalance of engine oil/gas ratio. The content of carbon monoxide in automobile exhaust gas does not cause death, but it occupies oxygen level, which can reduce the oxygen transfusion in blood, form anoxia, cause dizziness, nausea, headache and other symptoms, damage the central nervous system, and cause chronic poisoning.

2. Hydrocarbons (HC)

Hydrocarbons include unburned and incomplete burned fuels, lubricating oils and partial oxides, and contain aldehyde gases such as methane, formaldehyde and acrolein. Individual hydrocarbons are generally not very effective in humans, but they are an important part of photochemical smog production. When the concentration is high, it will have a strong irritating effect on eyes, respiratory tract and skin, and even cause dizziness, nausea, erythrocyte reduction, anemia and so on.

3. Nitrogen oxides (NO_x)

Nitrogen oxides are brown, pungent gases produced by engines. Nitrite and nitric acid are formed after nitrogen oxides enter human alveoli, which can stimulate lung tissue violently. Nitrite can combine with human body hemoglobin to form denatured hemoglobin, which can cause human body anoxia to a certain extent.

4. Particles (PM)

The harm of particles to human health is related to the size and composition of particles. The smaller the particles, the longer they stay in the air, and the greater the proportion of them stagnating in the lungs and bronchi after they enter the lungs, the greater the harm. In addition to being harmful to human respiratory system, particles have pores that can adhere to harmful substances such as SO_2, unburned HC and NO_2, thus causing greater harm to human health.

5. Photochemical smog

Nitrogen oxides photo chemically react with ultraviolet radiation in the sunlight, forming toxic photochemical smog, which is light blue and a secondary pollution of a highly irritating toxic gas. When the photochemical agent in the photochemical smog exceeds a certain concentration, it has obvious irritation, it can irritate the conjunctiva of the eye, cause tears and red eye disease, irritate the nose, pharynx and other organs, cause acute wheezing, make people breathe difficult, red throat pain, dizzy, and cause poisoning.

5.2.2 Air Quality Standard of Underground Parking Lot

According to *Indoor Air Quality Standard (GB/T 18883—2002)*, the air quality standard of underground parking lot is shown in Table 5-3.

Air quality standards Table 5-3

Pollutants	Carbon monoxide(CO) (mg/m³)	Hydrocarbon(HC) (mg/m³)	Nitrogen oxide(NO_x) (mg/m³)	Particulates(PM) (mg/m³)
Parameter	10 average value/h	0.61 average value/h	0.24 average value/h	0.15 average value/day

The start-up and acceleration process of the automobile in the underground parking lot are all idle. At idle speed, the ratio of emission of CO, HC and NO_x is about 7:1.5:0.2. Therefore, CO is the main harmful substance. As long as enough fresh air is provided to dilute CO concentration below the standard range, HC and NO_x can meet the standard requirements.

5.2.3 Ventilation of Underground Parking Lot

The ventilation methods of underground parking lot include natural ventilation and mechanical ventilation.

1. Natural ventilation

Because of the high noise in the start-up operation of underground parking lot ventilation system and the large amount of electric energy consumed in the operation, the frequency of mechanical ventilation system is not high. When designing mechanical ventilation system, natural ventilation measures should be considered as much as possible to achieve the purpose of energy conservation and environmental protection. The measures of natural ventilation mainly include the use of natural ventilation under the action of hot pressure and natural ventilation under the action of wind pressure.

1) Natural ventilation under hot pressure

Natural ventilation shafts may be provided. When a building is built on the ground parking lot, the auxiliary air duct shafts may be considered. Connect with the underground garage, open the hole and install the electric insulation valve to control the opening of the air valve. Outdoor air enters the opening of ventilation shaft of underground garage from door crack, ventilation shaft and lighting shaft of parking lot, and flows upwards along the auxiliary air duct shaft, and finally drains out outside. This is the natural ventilation under the action of hot pressure. The ventilation intensity is related to the height of the building and the temperature difference between the inside and outside.

2) Natural ventilation under wind pressure

Under the action of wind pressure, the building has the positive side of the wind pressure, and the negative side of the exhaust air, this is the natural ventilation under the action of wind pressure. The ventilation intensity is related to the opening area, the wind force of the positive pressure side and the negative pressure side. Therefore, in the design of underground garage, in the civil structure, we should design as many ventilation shafts as possible to increase the natural ventilation displacement under the action of wind pressure.

2. Mechanical ventilation

In recent years, ductless induced ventilation system has been widely used in underground parking lots, stadiums and other buildings in most cities of China. The ductless inductive ventilation system is composed of supply fan and exhaust fan. A large amount of air flow around that induction fan is induce and stirred by eject directional high-speed air flow from the nozzle of the induction fan, and the air in the garage is driven to flow along a pre-designed air flow to a target direction, namely, the directional air flow from the blower to the exhaust fan under the condition of no air pipe, so as to achieve the purpose of ventilation. Therefore, the ductless inductive ventilation system can effectively control the direction of air flow, The air inside the building is in a state of complete flow without stagnation dead angle of air flow, and the harmful substances are fully diluted and discharged from the outdoor by the exhaust fan, so that the comprehensive ventilation and ventilation inside the building can be realized, and the start-up and stop of the jet fan can be flexibly controlled, thereby achieving the effect of energy conservation.

5.2.4 Air Distribution in Underground Parking Lot

As for the air distribution of underground parking lot, two-thirds of the air volume is discharged from the upper and lower part of the parking lot, and one-third from the upper part of the parking lot. The air outlet shall be uniform and as close to the rear of the car as possible, so that the fume can not be collected and dispersed in any place. The main air outlet of the exhaust system shall be located at the top of the building or at the top of the apron far away from the main body to avoid secondary pollution, while the air outlet of the supply system shall be located at the main passageway to avoid short circuit between the supply and exhaust.

5.3 Mine Operation Ventilation

Mine ventilation is the most basic link in mine production, it always occupies a very important position in mine construction and production. The functions of mine ventilation are as follows:

(1) Supplying enough fresh air underground to meet the needs of personnel for oxygen;

(2) Diluting and removing toxic and harmful gas and dust in that well to ensure safe production;

(3) Adjusting the underground climate to create a good working environment;

(4) Improving the disaster resistance of the mine.

5.3.1 Composition and Properties of Air in Mine

Main air components in the mine:

1. Oxygen (O_2)

Oxygen is a colorless, odorless, odorless gas, its specific gravity to air is 1.11, its chemical properties are very active, and all gases can be combined, oxygen can help burn, oxygen is an indispensable metabolic substance for human and animals, without oxygen, people can not survive.

2. Carbon monoxide (CO)

Carbon monoxide is a colorless odorless gas with a specific gravity of 0.97 to air and is slightly soluble in water. The chemical properties of carbon monoxide are not active at normal temperature and pressure, but when the concentration reaches 13%-75%, the explosion can be caused by fire. Carbon monoxide is highly toxic because it has an affinity of 250-300 times greater than oxygen for the hemoglobin contained in human blood cells. Therefore, carbon monoxide inhaled into the human body will hinder the normal combination of oxygen and hemoglobin, resulting in hypoxia of tissues and cells in various parts of the human body, resulting in asphyxia and toxic death.

Source in the mine: blasting fume generated during blasting; exhaust gas from diesel engines; spontaneous combustion of coal seam, shale gas, etc.

3. Hydrogen sulfide (H_2S)

Hydrogen sulfide (H_2S) is a colorless, slightly sweet gas with odor of rotten eggs. It has a specific gravity of 1.19 to air, is soluble in water and can burn easily, and is explosive when the concentration reaches 4.3%-46%. It has very strong toxicity to cause blood poisoning and has severe irritation to eyes, mucous membrane and respiratory tract.

Source in the mine: decay of pit wood; sulfur-bearing minerals (e.g. pyrite, gypsum, etc.) decompose when saturated with water; emission from abandoned roadways in goaf or from coal surrounding rock; blasting fume.

4. Sulfur dioxide (SO_2)

Sulfur dioxide is a colorless gas with a strong sulphur burning smell. It has a specific gravity of 2.2 to air and is easily soluble in water. Often found at the bottom of the tunnel, it has a strong irritating effect on the eyes and respiratory organs. When the concentration of sulfur dioxide in the air was 0.0005%, the olfactory organs could smell the irritant odor. When its concentration was 0.002%, it had strong irritating smell, which could cause headache and sore throat. When its concentration is 0.05%, it can cause acute bronchitis and pulmonary edema, and die within a short time.

Source in the mine: spontaneous combustion or slow oxidation of sulphur-bearing minerals; releasing from coal surrounding rock; formed by explosion in sulphur minerals.

5. Nitrogen dioxide (NO_2)

Nitrogen dioxide is a reddish-brown gas, its specific gravity to air is 1.57, it is very easy to dissolve in water, to the eye nasal cavity, respiratory tract and lungs have a strong irritation, nitrogen dioxide combined with water to form nitric acid, so the lung tissue has a corrosive damage, can cause pulmonary edema.

Source in the mine: NO_2 is mainly produced by firing.

6. Hydrogen (H_2)

Hydrogen is colorless, tasteless and non-toxic, with a relative density of 0.07. The ignition temperature of hydrogen is 100-200℃ lower than that of methane. When the concentration of hydrogen in air is 4%-74%, it is dangerous to explode.

Source in the mine: H_2 is released from underground battery when charging; Hydrogen also gushes from some moderately metamorphic coal seams.

7. Ammonia (NH_3)

Ammonia is a colorless gas with strong odor. Its relative density is 0.596. It is easy to dissolve in water. When the concentration in air reaches 30%, it is dangerous to explode. Ammonia has irritating effect on skin and respiratory mucosa and can cause laryngeal edema.

Source in the mine: blasting work; gushing out of coal and rock strata, extinguishing fire with water, etc.

5.3.2 Air Concentration Regulations in the Mine

Safety regulations for metal and nonmetal mines (GB 16423—2006) has the following provisions for air in mines:

(1) The air composition (calculated by volume) in the intake air flow of the underground mining face shall not be less than 20% of oxygen and more than 0.5% of carbon dioxide;

(2) The dust content of the air source in the air entry shaft and working face shall not exceed 0.5 mg/m^3;

(3) The concentration of carbon monoxide in the air of the mine shall not exceed 30 mg/m^3;

(4) The concentration of sulfur dioxide in the air of the mine shall not exceed 15 mg/m^3;

(5) The concentration of nitrogen dioxide in the air of the mine shall not exceed 5 mg/m^3.

See Table 5-4 for the harmful gas concentration in the mine according to the *Coal Mine Safety Regulations*.

Hazardous gas concentration provisions   Table 5-4

Name of noxious gas	Symbol	Maximum allowable concentration (%)
Carbon monoxide	CO	0.0024
Nitrogen dioxide	NO_2	0.00025
Sulfur dioxide	SO_2	0.0005
Hydrogen sulfide	H_2S	0.0006
Ammonia	NH_3	0.004

5.3.3 Ventilation Method for Mine Excavation

A large number of lane ways must be dug in the process of mine construction, expansion, reconstruction or production. In order to supply fresh air, dilute and discharge harmful gases from coal (rock), blasting fume and mine dust, and create good weather conditions, ventilation must be carried out on the single heading face. This ventilation is called tunneling ventilation. The main ventilation methods include local fan ventilation, full air pressure ventilation and ejector ventilation.

1. Local fan ventilation

Local fan ventilation is the main method of local ventilation at present, which uses local fan as power and conducts air through the air tube. The common methods of local fan ventilation are press-in type, pull-out type and mixed type.

1) Pressed ventilation

Pressed ventilation is the use of local fans to push fresh air into the working face through the air tube, and flooding is discharged from the roadway, see Figure 5-2.

Chapter 5 Operation Ventilation for Other Underground Space

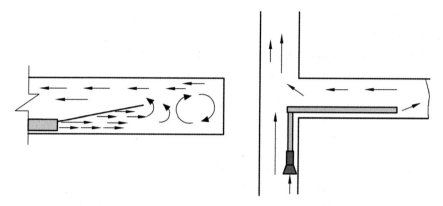

Figure 5-2 Schematic diagram of pressed ventilation

The local fan of pressed ventilation is installed in the fresh air flow, and the flooding air does not pass through the local fan, so that once the local fan has an electric spark, it is not easy to cause gas and coal dust explosion, so the safety is good, and the rigid air tube or the flexible air tube can be used, and the adaptability is strong. The disadvantage is that the working face is blown along the single roadway and drained back to the roadway, which is not conducive to the breathing of the workers in the roadway. After blasting, the speed of blasting fume exhausting from roadway is slow and the time is long, which affects the driving speed.

2) Exhaust ventilation

Contrary to the pressure ventilation, the fresh air enters the working face from the roadway, and the flooding air is discharged from the local fan through the air tube, see Figure 5-3.

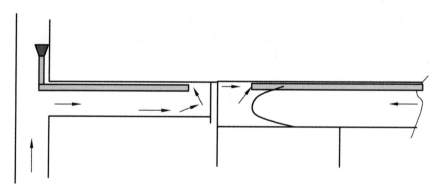

Figure 5-3 Exhaust ventilation schematics

Exhaust ventilation is beneficial to improve the speed of excavation because the polluted air is discharged through the air tube and the laneway is kept fresh air, so the sanitary conditions are good. The speed of fume exhausting after blasting is fast. But because the effective suction range at the end of the air duct is short, the air duct is easy to collapse when blasting. If the suction range is long, the ventilation effect is not good, and the safety of the polluted air passing through the local fan is poor. The hard air duct must be used in the exhaust ventilation, so the adaptability is poor.

3) Compound ventilation

Compound ventilation combines the two ventilations. Although overcoming some of the above shortcomings, but because of its many equipment, large power consumption, complex management, has not been promoted, see Figure 5-4.

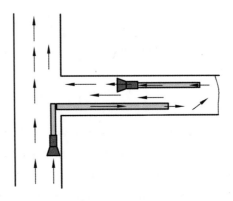

Figure 5-4　Compound ventilation schematics

2. Total pressure ventilation in mine

Total pressure ventilation is to introduce the fresh air of the main air flow into the driving face by means of the air pressure of the main ventilation fan in the mine and the air guide facilities. The ventilation rate depends on the available air pressure and wind resistance. Total pressure ventilation can be divided into:

1) Air guide by air duct

A windshield is arranged in the roadway to cut off the main air flow, the fresh air is introduced into the driving face with the air tube, and the polluted air is discharged from the blind heading driving roadway, see Figure 5-5.

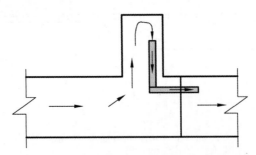

Figure 5-5　Schematic diagram of air guide by air duct

This method has the advantages of small auxiliary engineering quantity, convenient installation and disassembly of hairdryer, and is usually used in short roadway excavation ventilation with small air demand.

2) Air guide by parallel adit

At that same time of carry driving the main gallery, a ventilation gallery is parallel to the main gallery, and a connection channel is excavate between the main gallery and the distribution gallery at a certain distance to form a through air flow, and the old connection channel is closed when the new connection channel is communicated with the old connection channel, see Figure 5-6. Most of the single ends of the two parallel adits can be used for air reporting or air guide by blowers, while the rest of the gallery use the main gallery for air intake and connection channel for return.

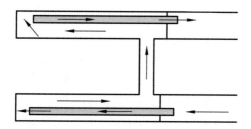

Figure 5-6 Schematic diagram of air guide in parallel adits

3) Air guide by windbreak

A longitudinal windbreak is arranged in that gallery, the fresh air on the upstream side of the windbreak is introduced into the excavation work surface, and the polluted air of the cleaning rear is discharged from the downstream side of the windbreak, see Figure 5-7.

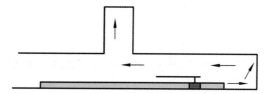

Figure 5-7 Schematic diagram of windbreak guide

The construction and removal of windbreaks by this wind guide method is a large amount of work. Suitable for short distances or when no other good method is available.

4) Ejector ventilation

Ejector ventilation is a method of ventilation of tunneling gallery (or other local work site) by using ejector. Its ventilation principle is to use pressurized water or compressed air through the nozzle high-speed jet. The surrounding air is sucked into the jet, mixed in the mixing tube, after obtaining energy, overcome the resistance of the air duct, move forward together, so that the air flow in the air duct continues to flow, to achieve the purpose of ventilation.

The advantages of this method are: no electrical equipment, no noise; but also has the functions of reducing temperature and dust; when the coal seam with serious coal and gas outburst is excavated, it is simple and safe to use instead of local ventilator. Disadvantages: low air pressure, small air volume, low efficiency, and the existence of gallery water problems.

5.4 Operation Ventilation of Civil Air Defense Works

Civil air defense works are generally confined underground space, many harmful gases are easy to accumulate in it. In this enclosed space, all kinds of building or decorative materials will emit some harmful gases. For example, some plastics or paint materials emit formaldehyde, while some rein supply concrete materials emit radon, and there are various pollutants or bacteria in the air, including even excessive carbon dioxide. Whether in times of war or in times of peace, it is certainly unhealthy for people to remain in a closed environment for long periods of time. Ventilation of civil air defense engineering is the precondition to ensure the air quality in the interior space of civil air defense engineering, which is very important in the design of civil air defense engineering.

5.4.1 Grade of Civil Air Defense Works

The protection category of civil air defense works is usually Class A and Class B civil air defense works.

The resistance levels of civil air defense engineering are usually: Constant 5, Nuclear 5, Nuclear 6, Nuclear 6B, etc.

5.4.2 Ventilation Design Parameters of Civil Air Defense Engineering

According to *Code for Design of Civil Air Defense Basement* (*GB 50038—2005*), the fresh air volume of personnel used in air defense basement in peacetime shall not be less than 30 $m^3/(p \cdot h)$ when ventilated, and the air conditioning shall conform to the provisions of Table 5-5.

Fresh air volume of air conditioner for personnel in peacetime [30 $m^3/(p.h)$]   Table 5-5

Room Functions	Fresh Air Volume of Air Conditioner
Hotel rooms, meeting rooms, hospital wards, beauty salon, recreation hall, dance hall, office	≥30
Restaurants, Reading Rooms, Libraries, Theatres, Shopping Malls (Stores)	≥20
Bar, teahouse, coffee shop	≥10

According to *Code for Design of Civil Air Defense Basement* (*GB 50038—2005*), the indoor air temperature and relative humidity of the air defense basement used at ordinary times shall be as specified in Table 5-6.

Chapter 5 Operation Ventilation for Other Underground Space

Indoor air temperature and humidity in peacetime   Table 5-6

Engineering and Room Categories	Summer		Winter	
	Temperature (℃)	Relative Humidity (%)	Temperature (℃)	Relative Humidity (%)
Hotel Rooms, Meeting Rooms, Offices, Multi-function Rooms, Library Reading Rooms, Entertainment Rooms, Wards, Shopping Malls, Theatres	≤28	≤75	≥16	≥30
Ballroom	≤26	≤70	≥18	≥30
Restaurant	≤28	≤80	≥16	≥30

According to *Code for Design of Civil Air Defense Basement* (*GB 50038—2005*), the wartime fresh air volume of the personnel in the air defense basement shall conform to the provisions of Table 5-7.

Fresh air volume for indoor personnel in Wartime $[30\ m^3/(p \cdot h)]$ Table 5-7

Air Defense Basement Category	Clean Ventilation	Toxic filtration and ventilation
Medical Emergency Project	≥12	≥5
Shelter and Production Space for Air Defense Specialists	≥10	≥5
Shelters for first-class personnel, food stations, regional water supply stations, power station control rooms	≥10	≥3
Second-class shelter	≥5	≥2
Other supporting projects	≥3	—

5.4.3 Ventilation for Civil Air Defense Works

The ventilation of civil air defense engineering can be divided into peacetime ventilation and wartime protective ventilation according to the ventilation opportunity.

1. Peacetime ventilation

The ventilation of civil air defense engineering in peacetime use or maintenance management is called peacetime ventilation. Peacetime ventilation is divided into normal ventilation and maintenance ventilation.

1) Normal ventilation at ordinary times

The combination of peacetime and wartime engineering, peacetime engineering normal use, a variety of ventilation and air conditioning equipment normal operation, this ventilation is known as peacetime normal ventilation. Because of the difference between peacetime function and wartime function, the protection equipment of ventilation system is in the state of waiting for installation or maintenance. In the face of war, we should take measures to change from peacetime to wartime.

2) Maintenance of ventilation at ordinary times

Wartime use, peacetime do not use the project, peacetime daily regular operation of a variety of ventilation and air conditioning equipment to complete maintenance and management work, this ventilation is known as peacetime maintenance ventilation.

2. Wartime protective ventilation

The ventilation of engineering in the face of war or in the state of war is called wartime protective ventilation. There are three kinds of protective ventilation in wartime, namely, clean ventilation, filter ventilation and isolated ventilation.

1) Clean ventilation

Clean ventilation means that the natural air outside the basement is not polluted, and it is a process that the air entering the basement is not polluted by any vent and need not be treated. This ventilation is consistent with the usual ventilation function, which is used to supplement fresh air for civil air defense project. In the civil air defense engineering, except for some equipment shelters, the natural ventilation of lanes is used to make up the air, and the mechanical ventilation is used to make up the air (the supply and exhaust fans are used in wartime).

2) Isolated ventilation

Isolated ventilation is a kind of ventilation in which the inlet and outlet passages of the project are completely closed and the air circulates inside the project. There are two operation methods of isolated ventilation: one is the single ventilation using the return air plug-in valve of the air intake system and the air intake fan. Second, the use of supply and return system of air conditioning ventilation. During the implementation of isolated ventilation, the protective door, protective closed door and closed door of the project mouth shall be closed, the explosion-proof wave valve of the project inlet and outlet, the closed valve of the inlet and outlet system and the automatic exhaust valve shall be closed. In the case of closed protection, personnel shall not enter or leave the project, nor shall they be allowed to supply water to the project or drain water from the project. The CO_2 concentration inside the project should also be detected during the airtight protection. When the CO_2 concentration exceeds the standard, the oxygen regeneration device should be activated.

3) Filter ventilation

The filtration ventilation is aimed at the outdoor air polluted by nuclear, biological and chemical warfare agents, which must be purified by dust removal and filtration facilities. There is an additional filter absorber in the air intake process and no exhaust fan in the air exhaust process, because to ensure the positive pressure in the civil air defense project from the penetration of toxic agents, the air flow can be guided to the air exhaust well through the overpressure exhaust valve and a series of wall-penetrating pipes arranged at the air exhaust port. Before underground works are transferred to filtration ventilation, it is necessary to find out the conditions of attacks outside the works and to find out the nature, types and concentrations of toxic agents.

The safe transition order of the three protective ventilations is: clean ventilation, isolated ventilation, filter ventilation (or clean ventilation). That is to say, when an enemy carries out an attack on a nuclear, chemical or biological weapon, he should immediately switch from clean

Chapter 5　Operation Ventilation for Other Underground Space

ventilation to isolated ventilation, and decide whether to switch to filter ventilation or clean ventilation after the chemical and biological prevention unit has ascertained the air pollution situation.

 Exercise

5.1　What are the ways for underground shopping malls to change the air sanitation environment?

5.2　Briefly describe the air treatment method of centralized all-air fixed air volume air conditioner.

5.3　What are the pollutants in the underground parking lot? What's the harm of being apart?

5.4　Describe the natural ventilation of underground parking lot.

5.5　Briefly describe the working principle of ductless induced ventilation system.

5.6　What are the main components of the air in the mine? What are the properties of each?

5.7　What are the ventilations for mine excavation? What are the advantages and disadvantages of each?

5.8　What are the ventilations for civil air defense works in wartime? What are the characteristics of each?

 Key vocabulary:

advection diffuser　平流散流器
air guided by air duct　风筒导风
air guided by parallel adit　平行巷道导风
air guided by wind break　风障导风
clean ventilation　清洁通风
ductless induced ventilation　无风道诱导型通风
ejector ventilation　引射器通风
filter ventilation　滤毒式通风
full air pressure ventilation　全压通风
humidifier　加湿器
intermediate filter　中效过滤器
isolated ventilation　隔绝通风
local fan ventilation　局部通风
louver outlet　百叶风口
mechanical fume exhaust　机械排烟
minimum fresh air volume　最小新风量
secondary heater　二次加热器
surface cooler　表面冷却器
supplier muffle　送风消音

Chapter 6   Disaster Prevention and Rescue of Highway Tunnels

[Important and Difficult Contents of this Chapter]
(1) Fire characteristics of highway tunnel, including tunnel area division, fire stage division, temperature change rule with time, temperature transverse distribution and temperature longitudinal distribution.
(2) Fire ventilation calculation and fume control standard in longitudinal ventilation.
(3) Formulation of disaster prevention and rescue plan for single tunnel.
(4) Formulation of disaster prevention and rescue plan for tunnel group.
(5) Classification of dangerous goods.

6.1   Fire Characteristics of Highway Tunnels

The following problems are studied by single tunnel fire experiment: tunnel area division, fire stage division, temperature variation with time, temperature transverse distribution and temperature longitudinal distribution.

1. Tunnel area division in fire

According to the flow state of fume and the pollution state of fume to tunnel, the tunnel in fire can be divided into three areas.

1) The upstream of fire area

This area is located on the upwind side of the combustion area. When the wind speed is high, the area is not polluted by the fume, and the air flow structure and gas composition are not affected by the fire. However, due to the dynamic effect of the fire and the change of ventilation system caused by evacuation and rescue, the air flow velocity, temperature, density and static pressure in the upper reaches of the fire area have some changes. When the control is good, the area is a safe passage for firefighting and rescue.

2) Fire area

The area is a collection of areas in which combustibles are burned and areas in which the flame has reached. On the one hand, the flame occupies a part of the tunnel section, reducing the cross-sectional area of the overflow, thus increasing the ventilation resistance of the section. On the other hand, because of the increase of the fire temperature and the expansion of the gas, the resistance of the section also changed significantly. As that volume of the gas expand, however, the volume flow rate will increase. As a result of that combustion of the combustible material in the fire area, the composition of the gas will change greatly, and the temperature will also increase greatly, see Figure 6-1.

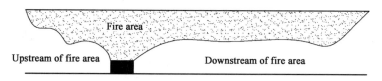

Figure 6-1 Division of tunnel area in fire

3) The downstream of fire area

This area is located in all fume-contaminated areas on the downwind side of the fire area. The air flow state of the tunnel is significantly different from that before the fire. The fume temperature is increased, the density is decreased, the concentration of toxic and harmful gases is increased, and the resistance coefficient of the tunnel is changed. In the process of flue gas flow, its temperature, density, velocity, gas composition and concentration change constantly.

2. Fire stage division

According to the change of fume temperature with time, the tunnel fire process can be divided into three stages.

1) Fire development stage

In this stage, the oxygen supply is sufficient, generally oxygen-rich combustion, the fire is increasing, the maximum temperature of the flue gas is increasing, the concentration of oxygen in the flue gas is decreasing, the concentration of carbon dioxide, carbon monoxide, hydrogen, methane and other toxic gases is increasing. In the development stage, the combustion state is unstable, and the air flow state is unstable, so the change rule of the flue gas temperature along the flue gas is not obvious with the maximum temperature of the flue gas.

2) Fire stabilization stage

At this stage, the fire is basically stable, the maximum temperature of fume flow changes little, if the fuel is sufficient and there is supply, the fire will continue. With the increase of the distance from the fire area, the temperature of the flue gas decreases. With the increase of fire time, the temperature gradient of flue gas increases in the tunnel near the fire area, while decreases in the tunnel far away from the fire area with the increase of fire time.

3) Fire attenuation stage

In this stage, the fire gradually decreased, the maximum temperature of the fume decreased with the increase of time, the concentration of oxygen in the fume increased, the concentration of carbon dioxide and carbon monoxide decreased. With the increase of distance from the fire area and the increase of fire time, the temperature gradient of the fume flow and the temperature gradient of the fume flow along the fire area decrease.

3. Law of temperature change with time

When a fire occurs in the tunnel, the combustion starts from a point or a surface, and with the time delay, the flame spreads to the surrounding area, and the fire area burns into an area. With the gradual increase of the amount of combustibles involved in the combustion, the heat increases rapidly, and the temperature of the fire fume increases.

Through the fire experiment, the fire process curve of the fire area under different wind speed is obtained, see Figure 6-2.

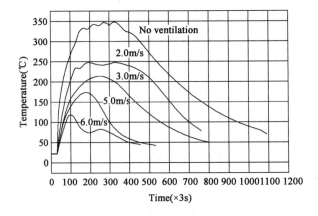

Figure 6-2 Process curves of fire with different wind speed

Through comparative analysis, it can be found that the change of temperature in the tunnel does not increase gradually according to the standard temperature-time curve, but has a sharp increase process. Generally, the temperature reaches the highest within 2-10 minutes after the fire. Among them, the highest temperature is related to the type and quantity of burning materials, the duration of burning, the burning rate and the characteristics of the tunnel itself.

When the fire enters the stable combustion stage, its duration varies with the size of the fire, ventilation speed and the free surface area of the fuel. Under the same conditions, the larger the fire scale, the longer the duration of the fire; the larger the free surface area of the fuel and the faster the burning rate, the shorter the duration of the fire. At the same time, under the same conditions, with the increase of ventilation speed, the duration of fire is shortened.

4. Transverse distribution of temperature

With no ventilation, the upper part of the tunnel cross-section is a turbulent layer in which high-temperature fume flows to both ends of the tunnel, while the lower part of the tunnel cross-section is a turbulent layer in which fresh air from outside flows into the tunnel to supplement, and the middle part of the tunnel cross-section is a turbulent layer in which high-low temperature air conducts heat and convects. Because the high temperature fume flow is lighter and rises, the tunnel bottom has a relatively cold air supplement, so the transverse distribution of temperature is the highest in the vault, the second in the arched waist and the side wall, and the lowest in the bottom. However, the maximum temperature is not on the surface of the vault lining, but in the region not far from the vault, which is due to the heat exchange between the high-temperature fume and the lower-temperature lining in the region close to the vault, so the temperature of the fume in this region is slightly lower than the temperature of the fume at a certain distance from the vault.

Under the condition of mechanical ventilation, when the ventilation speed is ⩾ 2.0 m/s, because the flame is blown obliquely and depressed, the bottom of the section near the downstream of the fire area is grilled by the flame, the temperature rises sharply, the cross-sectional temperature distribution shows the rule of bottom high and vault low. With the distance away from the fire area, due to the rising of high-temperature fume, the cross-sectional temperature distribution becomes the highest at the vault, followed by the arched waist and side wall, and the lowest at the bottom. At the same time, with the increase of the distance to the fire area, the temperature distribution on the cross-section becomes more and more uniform.

5. Longitudinal distribution of temperature

Under the combined action of mechanical ventilation power, natural air pressure and fire-air pressure, high-temperature fume flows to the downwind side of the fire area, and its influence area gradually expands with the passage of time. At that same time, because the fume flow temperature is high than the lining temperature of the tunnel along the road, in the diffusion process, the fume flow continuously carries on the heat exchange with the surrounding lining and other object, the fume loses heat, the temperature drops gradually, the tunnel lining gets heat, the temperature rises unceasingly.

The relationship between ventilation speed, fire scale and the longitudinal distribution of temperature field in the tunnel was obtained by fire test, see Figure 6-3.

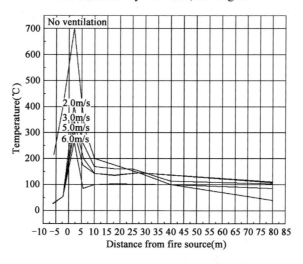

Figure 6-3 Longitudinal temperature distribution on the vault

Based on the fire test and theoretical analysis, it is known that the heat of high temperature fume is absorbed by air and non-combustible materials (mainly tunnel wall) and the temperature of high temperature fume decreases along the tunnel due to the convection of hot and cold air caused by combustion and the cooling effect of tunnel wall on the high temperature fume flowing there through. Therefore, the distribution law of the temperature field in the longitudinal section of the tunnel is: with the increase of the distance to the fire point, the temperature decreases, and the temperature gradient gradually decreases.

6.2 Standard for Calculation of Fire Ventilation Pressure and Fume Control in Highway Tunnels

At present, longitudinal ventilation is generally used in highway tunnels, and longitudinal ventilation can be used for extra-long tunnels. This section mainly discusses the standard for calculation of fire ventilation pressure and fume control in longitudinal ventilation.

6.2.1 Calculation of Ventilation Pressure

The composition of ventilation pressure in highway tunnel under various working conditions is in Table 6-1.

Ventilation Pressure under Various Working Conditions   Table 6-1

Number	Ventilation Pressure	Normal Operation Ventilation	Maintenance Ventilation	Fire Ventilation
1	Throttling Effect Fume Resistance			√
2	Fire and Wind Pressure			√
3	Blocking Section Resistance			√
4	Fume Friction Resistance	√	√	√
5	General Friction Resistance	√	√	√
6	Local Resistance	√	√	√
7	Lifting pressure of Jet Fan	√	√	√
8	Lifting pressure of Axial Fan	√	√	√
9	Traffic Wind Pressure	√		√
10	Natural Wind Pressure	√	√	√
11	Lifting pressure of Mobile Fan		√	√

1. Throttling effect fume resistance

When the change of the molar mass of the fire fume is ignored, the resistance of the throttling effect fume can be calculated by Formula (6-1):

$$H_j = \frac{1}{2}\rho_1 \left[v_1^2 \left(\frac{1}{M_k} - 1 \right) + gh_m \cos\beta (1 - M_k) \right] \quad (6-1)$$

Where: ρ_1——Density of air flow before fire (kg/m^3);

Chapter 6 Disaster Prevention and Rescue of Highway Tunnels

v_1——Velocity of air flow before fire(m/s);

M_k——Relative change of the products of fire combustion;

h_m——Tunnel height(m);

β——Tunnel slope;

g——Gravitational acceleration(m/s^2).

2. Fire and Wind Pressure

Assuming that the air pressure at each point remains unchanged after the fume flows through the tunnel, the fire and wind pressure can be calculated according to Formula (6-2):

$$h_b = 11.77\Delta Z \frac{\Delta t}{T} \tag{6-2}$$

Where: ΔZ——Elevation difference of high temperature gas flowing through the tunnel(m);

Δt——Increase in the average temperature of the air flowing through the tunnel by high-temperature gas (K);

T——The mean absolute temperature of the air after a fire in the tunnel caused by high-temperature gases (K).

For long highway tunnels, because of the change of temperature along the road, when calculating the fire and wind pressure, the tunnel under fire can be divided into sections, the fire and wind pressure of each section can be calculated separately, and then superimposed. As shown in Formula (6-3):

$$h_b = \Sigma h_{bi} = \Sigma 11.77\Delta Z_i \frac{\Delta t_i}{T_i} \tag{6-3}$$

Where: ΔZ_i——Elevation difference of the tunnel in section i(m);

Δt_i——Increment of mean air temperature in the tunnel of section i (K);

T_i——Mean absolute temperature of the air after a fire in the tunnel in section i(K).

After a fire, the fume temperature increment is estimated by Formula (6-4):

$$\Delta T = \Delta t_0 e^{-\frac{c}{g}x} \tag{6-4}$$

Where: x——Distance from the flue gas temperature rise point to the fire source point calculated along the flue gas direction (m);

ΔT——Calculate the temperature increment at the distance of x m from the fire source along the direction of the temperature current(℃);

Δt_0——Temperature increment at the ignition point before and after ignition(℃);

g——Gravity acceleration(m/s^2);

c——Coefficient, $c = kU/3600C_p$;

k——Rock thermal conductivity, $k = 2 + k'\sqrt{v}$, k' value is range from 5 to 10;

U——Perimeter of the roadway at the corresponding calculated point(m);

C_p——Specific heat of air at constant pressure, 0.2 kcal/(kg/℃).

3. Blocking section resistance

The blocking section resistance can be calculated according to Formula (6-5):

$$P'_f = (R'_\lambda + R'_\xi)Q^2 = \left(\lambda_h \frac{l_T \rho S'}{8F'^3} + \xi' \frac{\rho}{2}\frac{1}{F'^2}\right)Q^2 \tag{6-5}$$

Where: λ_h——Coefficient of drag along the inverted annular space airflow rack; ;

l_T——Blocking section length(m);

Q——Air volume of the blocking section (m³/s);

ρ——Air density(kg/m³);

ξ'——Local resistance coefficient of reverse blocking section;

S'——Wet perimeter of cross section of reverse blocking section(m);

F'——Cross section area of the blocking section(m²).

4. Fume friction resistance

Fume friction resistance can be calculated according to Formula (6-6):

$$h_f = \sum \frac{R_a S^2 \rho v^2}{\rho_a} L \tag{6-6}$$

Where: S——Cross-sectional area of the inverted side tunnel(m²);

ρ——Air density(kg/m³);

v——Velocity of flue gas flow(m/s);

ρ_a——Density of air flow rack before fire(kg/m³);

L——Tunnel length(m);

R_a——Wind resistance per unit length of tunnel before fire(Ns²/m⁸).

The wind resistance R_a per unit length of the tunnel before the fire can be calculated according to Formula (6-7):

$$R_a = \frac{\lambda \rho_a U}{8S^3} \tag{6-7}$$

Where: λ——Tunnel friction coefficient;

U——Perimeter of tunnel cross section(m).

For long highway tunnels, because of the change of fume flow along the road, when calculating fume flow resistance, the tunnel under fire can be divided into sections, the fume flow resistance values of each section can be calculated separately, and then superimposed. As shown in Formula (6-8):

$$h_f = \sum h_{fi} = \sum \frac{R_{ai} S_i^2 \rho_i v_i^2}{\rho_{ai}} L_i \tag{6-8}$$

Where: S_i——Cross-sectional area of the tunnel in the section i(m²);

ρ_i——Air density in tunnel section i(kg/m³);

v_i——Fume flow velocity of the tunnel in the section i(m/s);

ρ——Density of air flow before fire of the tunnel section i(kg/m³);

L_i——Length of tunnel section i(m);

R_{ai}——Wind resistance per unit length of tunnel before fire in tunnel section i(Ns²/m⁸).

5. General frictional resistance

When the ventilated air flows in the tunnel and the wellbore at the speed u_r, the resistance due to the friction of the wall surface (hereinafter referred to as the friction resistance) can be calculated according to Formula (6-9):

$$\Delta P_\lambda = \lambda \frac{L}{D} \frac{1}{2} \rho u_r^2 \qquad (6-9)$$

Where: λ——Tunnel friction coefficient;
$\quad\rho$——Air density in the tunnel(kg/m^3);
$\quad u_r$——Velocity of air flow in the tunnel(m/s);
$\quad L$——Tunnel length(m);
$\quad D$——Equivalent diameter of tunnel clearance section(m).

6. Local resistance

The pressure loss caused by the sudden expansion or contraction of air flow, turning, crossover, convergence and other conditions, the magnitude and direction of the wind speed will change, which is called local resistance and can be calculated according to Formula (6-10):

$$\Delta P_\xi = \xi \frac{1}{2} \rho v^2 \qquad (6-10)$$

Where: ξ——Local resistance coefficient;
$\quad\rho$——Air density in the tunnel(kg/m^3);
$\quad v$——Air speed in the tunnel(m/s).

7. Lifting pressure of jet fan

The lifting pressure of each jet fan shall be calculated according to Formula (6-11):

$$P_j = \rho v_j^2 \frac{A_j}{A_i}\left(1 - \frac{v_r}{v_j}\right)\eta \qquad (6-11)$$

Where: P_j——Lifting pressure of a single jet fan(N/m^2);
$\quad v_j$——Air speed of jet fan outlet(m/s);
$\quad v_r$——Design wind speed in tunnel(m/s);
$\quad A_j$——Area of air outlet of jet fan(m^2);
$\quad A_i$——Area of tunnel clearance section(m^2);
$\quad \eta$——Reduction factor of friction loss at the position of jet fan, taken as 1.1.

Assuming that there are n_j jet fans, the total lifting pressure of jet fans is as follows:

$$\Delta P_j = n_j P_j \qquad (6-12)$$

8. Lifting pressure of axial fan

Lifting pressure of each axial fan can be calculated according to Formula (6-13):

$$P_z = aq^2 + bq + c \qquad (6-13)$$

Where: P_z——Lifting pressure of axial fan(N/m^2);
$\quad q$——Blowing rate of axial fan(m^3/s);

a,b,c——Parameters of fan characteristic curve, which can be solved by Lagrange quadratic interpolation method or least square method.

9. Traffic wind pressure

The calculation of tunnel ventilation must be based on the analysis of the ventilation system of a specific project. In the case of traffic jam or two-way traffic, traffic ventilation should be considered as resistance. In the normal operation of one-way traffic, the traffic ventilation wind should be considered as the power, but when the vehicle speed is less than the design wind speed, the traffic ventilation wind should be considered as the resistance.

The wind pressure ΔP_i generated by the traffic wind in the single-hole two-way traffic tunnel can be calculated according to Formula (6-14):

$$\Delta P_t = \frac{A_m}{A_r} \frac{\rho}{2} n_+ (v_{t(+)} - v_r)^2 - \frac{A_m}{A_r} \frac{\rho}{2} n_- (v_{t(-)} - v_r)^2 \quad (6\text{-}14)$$

Where: A_m——Equivalent impedance area of vehicle (m^2);

v_r——Design wind speed of tunnel (m/s), $v_r = \frac{Q_r}{A}$;

Q_r——Design wind volume of tunnel (m^3/s);

n_+——Number of vehicles whose driving direction are same with the wind flowing direction (veh), $n_+ = \frac{N_+ L}{3600 v_{t(+)}}$;

n_-——Number of vehicles whose driving direction are opposite with the wind flowing direction (veh), $n_- = \frac{N_- L}{3600 v_{t(-)}}$;

N_+——Design peak hour traffic volume of tunnel (in the same direction with wind blowing) (veh/h);

N_-——Design peak hour traffic volume of tunnel (in the opposite direction with wind blowing) (veh/h);

$v_{t(+)}$——Speed of vehicles whose driving direction are same with the wind flowing direction (m/s);

$v_{t(-)}$——Speed of vehicles whose driving direction are opposite with the wind flowing direction (m/s).

The wind pressure generated by traffic wind in one-way traffic tunnel can be calculated according to Formula (6-15):

$$\Delta P_t = \frac{A_m}{A_r} \frac{\rho}{2} n_C (v_t - v_r)^2 \quad (6\text{-}15)$$

Where: n_C——Number of vehicles in the tunnel, $n_C = \frac{NL}{3600 v_t}$;

v_t——Vehicle speed under various operation conditions (m/s).

10. Natural wind pressure

When the wind pressure generated by the natural wind coincides with the direction of the traffic wind, the thrust force is generated, and when the wind pressure generated by the natural wind coincides with the direction of the traffic wind, the resistance force is generated. The value can be calculated according to Formula (6-16):

$$\Delta P_m = \left(\xi_e + \xi_0 + \lambda \frac{L}{D_r}\right)\frac{\rho}{2}v_n^2 \qquad (6\text{-}16)$$

Where: ξ_e——Local resistance coefficient at tunnel entrance;
ξ_0——Local resistance coefficient at tunnel exit;
λ——Coefficient of frictional resistance related to the relative roughness of the tunnel lining surface;
L——Tunnel length(m);
D_r——Equivalent diameter of tunnel clearance section(m);
v_n——Natural wind speed in tunnel(m/s)。

6.2.2 Standard for Fume Control

Generally speaking, in the case of fire, from the safety point of view, the following principles should be adopted in the design: control the spread of fume, and make people evacuate in fume-free state as far as possible. In all cases, people must be able to reach a safe place in a reasonably short time and a reasonably short distance by providing facilities such as emergency exits or fire barriers. Ventilation systems must be able to ensure that escape routes and standby points are fumeless. The ventilation system must be able to create good conditions for firefighting. In the case of gasoline combustion, indirect explosions due to incomplete combustion must be avoided. Therefore, the ventilation system must be able to provide sufficient air to allow it to burn fully or to dilute explosive gases. In order to minimize the area of the fire where fuel vaporization occurs, an appropriate drainage system should be provided.

There are three scenarios of highway tunnel fire: fire in single tunnel, fire in double tunnel, fire in single tunnel combined with ventilation shaft. For different fire scenarios, the corresponding rescue methods are different, so the ventilation and fume control standards are different.

1. Standard for fume control in single tunnel fire scenarios

When a fire occurs in the tunnel, a large number of high-temperature fume will be generated. Under the condition of high ventilation speed, the high-temperature fume will fill the whole tunnel section in a very short time (20-30s), and the visibility in the tunnel will be reduced to about 1 m. Generally speaking, once a fire occurs in a single tunnel, the only feasible way to clear the fume is to exhaust the fume from the exit of the tunnel longitudinally. At this time, the tunnel needs to be ventilated longitudinally. When the ventilation speed is high, it will cause eddy current and destroy the stratified structure of the fume flow. The higher the ventilation speed, the more obvious this phenomenon is. Therefore, it is necessary to control the ventilation speed in the tunnel within a reasonable range in case of fire.

When the ventilation speed in the tunnel is too low and the fire scale is too large, the stratified flow of fume flow and air flow will occur in the fire process. In the vicinity of the fire area, fume flows along the roof against the wind for a period of time, a phenomenon known as fume countercurrent (Figure 6-4). When the ventilation speed in the tunnel is too high, on the one hand, the fire will spread to the downstream speed will be accelerated, on the other hand, the lift force generated by the buoyancy action will not be able to drive the fume to flow upwards. Downstream of the fire, the fume is in a turbulent state, and the fume appears the phenomenon of bottom layer, that is, the fume mainly flows in the lower part of the tunnel (Figure 6-5).

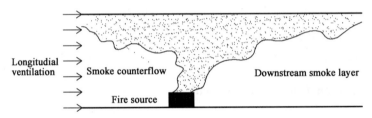

Figure 6-4　Fume countercurrent with too low ventilation speed

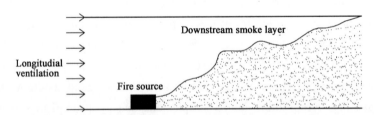

Figure 6-5　Fume in the bottom with too high ventilation speed

Through the above analysis, we can see that there is a critical wind speed. Under this wind speed, there will be no fume counter-current in the tunnel and the diffusion speed to the downstream is not very fast. At the same time, there will be no fume in the bottom. At this time, there is clean and breathable air below the fume layer downstream of the fire, which brings convenience to evacuation and rescue.

According to the formula recommended by PIARC, the critical ventilation rate can be approximated by Formula (6-17):

$$v_c = K_1 K_2 \left[\frac{gHQ}{\rho C_p A v_c} + T \right]^{1/3} \quad (6\text{-}17)$$

Where: K_1, K_2——Constants;

　　　　g——Gravity acceleration (m/s^2);

　　　　H——Tunnel height (m);

　　　　Q——Heat release rate of fire field (MW);

Chapter 6 Disaster Prevention and Rescue of Highway Tunnels

A——Cross-sectional area of the tunnel(m^2);
C_p——Specific heat of air, 0.2kcal/(kg℃);
ρ——Ambient air density(kg/m^3);
T——Ambient air temperature(℃).

Critical wind speeds can also be determined based on three-dimensional computer numerical simulation analysis, taking into account such parameters as the width of the fire site (Figure 6-6).

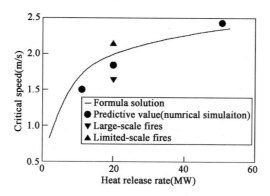

Figure 6-6 Calculated and theoretical values of critical wind speed

Under general conditions, in single tunnel fire, the longitudinal wind speed in the tunnel is controlled at 2-3 m/s, which can avoid the phenomenon of backflow and prevent the fire scope from expanding. Therefore, the fume control standard for single tunnel fire scenarios is: the direction of air flow is longitudinal, from the upstream of the fire area to the downstream of the fire area, and the wind speed is controlled at 2-3 m/s.

2. Standard for fume control in dual tunnel fire scenarios

Under the condition of double tunnels, across channel is set up between the two tunnels at intervals of a certain distance. The cross channel is used as the evacuation and rescue channel of the tunnel, also known as the connection channel, which can be divided into two types: the cross channel of vehicles and the cross channel of pedestrians. The transverse passageway is mainly used to evacuate vehicles, and the spacing is large. Pedestrian crossings are emergency passages for pedestrians to escape safely, and fire and rescue workers can also use the crossings to reach the fire site.

When a tunnel fires between two cross channels (Figure 6-7), the fire upstream cross channels need to be opened for evacuation and rescue, and the fire downstream cross channels should be closed. At this time, in order to ensure the safety of evacuation and rescue, we should ensure:

(1) Fume will not pollute the non-fire tunnel from the fire tunnel through the upstream cross channel. In order to achieve this goal, the wind speed of the upstream cross channel near the fire point should flow from the non-fire tunnel to the fire tunnel.

(2) Prevent backflow of fume from igniting fire.

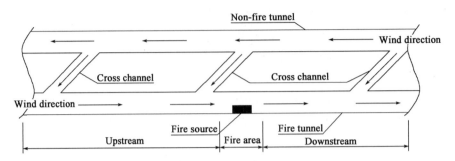

Figure 6-7 Ventilation and fume control in fire of double tunnel

(3) In order to ensure the smooth evacuation of personnel and vehicles, the wind speed in the cross channel should be controlled within a certain range.

Based on the experimental study of various ventilation combinations, for tunnels with cross channels, When the ventilation speed in the fire tunnel is 3 m/s and that in the non-fire tunnel is 5 m/s, and the direction of the air flow in the two cross channels upstream of the fire point flows from the non-fire tunnel to the fire tunnel, the flue gas backflow will not occur in the fire tunnel, and the high-temperature flue gas will not flow from the fire tunnel to the non-fire tunnel through the cross channels. Therefore, the fume control standard for the fire scene of the double tunnel with multi-cross channels is: the direction of the air flow in the fire tunnel is longitudinal, the direction of the air flow from the upstream of the fire area to the downstream of the fire area, and the wind speed is controlled at 3 m/s; The direction of air flow in the non-fire tunnel is longitudinal, and the wind speed is controlled at 5 m/s, which is opposite to that in the fire tunnel. The direction of air flow in the two cross channels upstream of the fire point is from the non-fire tunnel to the fire tunnel.

3. Standard for ventilation and fume control in fire scenario in single tunnel with ventilation shaft

In the ventilation of extra-long tunnels, it is generally necessary to adopt the longitudinal ventilation with ventilation shafts. In the actual situation, some tunnel shafts are hundreds of meters high, so the chimney effect is very obvious. The fire wind pressure has a great impact on the temperature field of tunnel fire, fume diffusion and so on. Therefore, it is necessary to give the standard for ventilation and fume control in fire occurring in different positions of the shaft supply and exhaust tunnel.

1) Fire occurs in the exhaust section

When the fire occurs in the exhaust section, most of the high-temperature fume produced by the combustion should be discharged from the exhaust shaft to the tunnel, so as to avoid the spread of high-temperature fume to the short channel between the vertical shafts, reduce the influence of the fire on the supply section, and restrain the occurrence of flue gas backflow.

Through the simulation experiment and calculation, the reasonable ventilation organization is obtained as follows:

Before personnel evacuation is completed, the wind speed combination of the supply-exhaust shaft is 0-3 m/s (Figure 6-8).

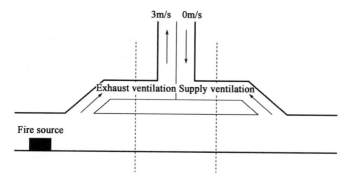

Figure 6-8   Before evacuation

After evacuation, the wind speed combination of the supply-exhaust shaft is 6-3 m/s (Figure 6-9).

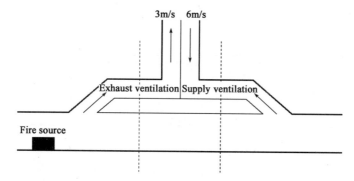

Figure 6-9   After evacuation

2) Fire occurs in vertical shaft short track

As that fire occur in the short passage between the vertical shafts, the reasonable ventilation organization should be carried out so that most of the high-temperature fume generated by the combustion is discharged from the vertical shafts, the flow of the high-temperature fume to the exhaust air and the spread of the supply section can be avoided, and the influence range of the high-temperature toxic fume flow can be reduced.

Through the simulation experiment and calculation, the reasonable ventilation organization is obtained as follows:

Before personnel evacuation is completed, the wind speed combination of the supply-exhaust vertical shaft is 0-3 m/s (Figure 6-10).

After evacuation, the wind speed combination of the supply-exhaust vertical shaft is 3-3 m/s (Figure 6-11).

3) Fire occurs in the supply section

When the fire occurs in the supply section, the high-temperature fume produced by the combustion shall be exhausted from the supply section as far as possible to prevent the fire from spreading to the short tunnel between vertical shafts.

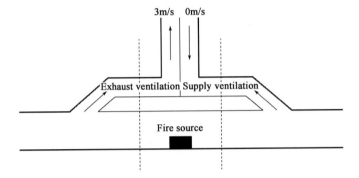

Figure 6-10   Before evacuation

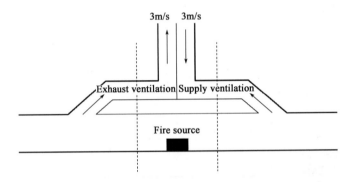

Figure 6-11   After evacuation

Through the simulation experiment and calculation, the reasonable ventilation organization is obtained as follows:

Before personnel evacuation is completed, the wind speed combination of the supply-exhaust vertical shaft is 0-3 m/s (Figure 6-12). After evacuation, the wind speed combination of the supply-exhaust vertical shaft are shown in Figure 6-13.

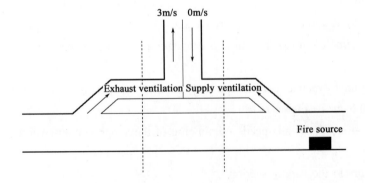

Figure 6-12   Before evacuation

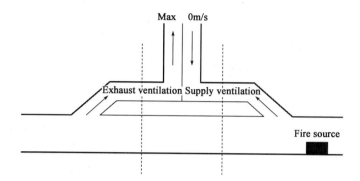

Figure 6-13   After evacuation

6.3   Disaster Prevention and Rescue System for Single Tunnel

In the design of disaster prevention and rescue of highway tunnel, the following general principles should be implemented: people-oriented, prevention-oriented and combination of prevention and elimination; effective monitoring, effective measures, orderly evacuation, combined rescue and self-rescue; early detection, timely fire suppression, mobile and fixed fire suppression combined.

(1) People-oriented, prevention-oriented and combination of prevention and elimination. The highest principle is to minimize the hazards of fire to the tunnel personnel, and the driving safety guarantee system for the detection and management of hidden fire hazards is established, as well as the fire alarm, rescue and fire extinguishing prevention system. The following considerations shall be made in terms of hardware and software.

The software includes: normal driving rules and regulations, treatment plan for non-fire abnormal situation, measures for the inspection and traffic control of vehicles carrying dangerous goods, daily monitoring and management system, alarm and fire control system inspection and maintenance system.

Hardware includes: equipment and cable fire-resistant design, layout spacing of cross channels of vehicle and pedestrian and connection method with main tunnel, equipment arrangement, monitor and alarm system, firefighting equipment, dangerous goods vehicle inspection equipment, etc.

(2) Effective monitoring, effective measures, orderly evacuation, combination of rescue and self-rescue: Establishment of high-standard facilities and methods for personnel rescue and self-rescue; The following considerations should be taken into account in terms of hardware and software:

The software includes: organization and execution plan in case of fire, ventilation plan in case of fire, Traffic organization plan in case of fire, evacuation and rescue plan in case of fire, a fire-fighting plan in the event of a fire.

Hardware includes: alarm facilities, escape channel signs, guidance facilities, self-rescue equipment, rescue facilities, fire-fighting facilities.

(3) Combining early detection, timely fire extinguishing, mobile fire extinguishing with fixed fire extinguishing: focusing on early fire extinguishing, extinguishing the fire in 5-10 minutes before deflagration to minimize the loss; The following considerations should be taken into account in terms of hardware and software:

Software includes: setting up early fire extinguishing and deflagration extinguishing plans, fire extinguishing and ventilation organization plan, fire-fighting method plan and work plan of fire-fighting team, fire drills.

Hardware includes: intelligent mobile fire extinguishing facilities, fixed fire extinguishing facilities and so on.

In order to realize the general principles of disaster prevention and rescue in highway tunnels, the following work must be done:

(1) Divide the grade of highway tunnel traffic engineering.

(2) According to the grade of tunnel traffic engineering, determine the configuration type of tunnel traffic engineering facilities.

(3) According to the configuration type of tunnel traffic engineering facilities, combined with the tunnel civil engineering design conditions, the tunnel monitoring system is designed.

(4) According to the design documents of the tunnel monitoring system, the addresses of all kinds of facilities are coded.

(5) Divide the tunnel fire prevention, and determine the operation equipment of each system and its corresponding address code in case of fire in each fire prevention division.

(6) Prepare disaster prevention and rescue plans according to the operation equipment of each system and its corresponding address code in the case of fire in each fire prevention area.

6.3.1 Grade Classification of Tunnel Traffic Engineering

The traffic engineering of highway tunnels has four grades: A, B, C and D according to the tunnel length and traffic volume. Highway tunnel traffic engineering grade can be calculated according to Formula (6-18):

$$P = L \times q \times 365 \times 10^{-10} \tag{6-18}$$

Where: P——Estimated value of the annual accident probability in the tunnel (When $P > 1$, taken as 1);

L——Tunnel length (m);

q——Annual average daily traffic volume for single hole design of tunnel (pcu/d).

According to the calculated value of P, the traffic engineering grade of the tunnel is divided as shown in Table 6-2.

Grade classification of traffic engineering of highway tunnels　　Table 6-2

P	Grade
P > 0.55	A
0.55 ≥ P ≥ 0.18	B
0.18 > P > 0.05	C
P ≤ 0.55	D

6.3.2 Allocation Standard of Tunnel Traffic Engineering Facilities

Classification of tunnel traffic engineering level Tunnel traffic engineering facilities allocation is generally based on the tunnel traffic engineering level allocation for special tunnels can be up to the level of one. See Table 6-3 for the allocation standard of tunnel traffic engineering facilities.

Tunnel traffic engineering facilities allocation standards　　Table 6-3

System	Facility	Tunnel Grade			
		A	B	C	D
Traffic Safety Facility	Tunnel Indicator	●	●	●	●
	Evacuation Indicators	●	■	🖳	🖳
	Emergency Telephone Indicator	●	■	🖳	🖳
	Fire hydrant Indicator	●	■	🖳	🖳
	Pedestrian Cross channel Indicator	●	■	🖳	🖳
	Vehicles Cross channel Indicator	●	■	🖳	🖳
	Marking Line	●	●	●	●
	Contour Mark	●	●	●	●
	Raised Road Sign	●	●	●	●
Traffic Monitoring System	Traffic Detector	●	■	■	
	Camera	●	■	■	
	Traffic Area Control Unit	●	■	■	
	Variable Speed Limit Sign	●	■	🖳	
	Variable Information Board	●	■	🖳	
	Traffic Light	●	■	■	🖳
	Lane Indicator	●	■	■	

continued

System	Facility	Tunnel Grade			
		A	B	C	D
Ventilation and Lighting Control	VIDetector	●	■	🗄	
	CODetector	●	■	🗄	
	NOXDetector	●	🗄	🗄	
	Wind Direction Anemometer	●	🗄	🗄	
	Ventilation Area Control Unit	●	■	🗄	
	Brightness Detector	●	■	🗄	
	Lighting Area Control Unit	●	■	🗄	
Connecting System	Emergency Call	●	●	🗄	
	Cable Broadcast	●	■	🗄	
Fire-fighting System	Fire Detector	●	■	🗄	
	Alarm Button	●	■	🗄	🗄
	Fire Control Machine	●	■	🗄	
	Fire Extinguisher	●	●	●	●
	Fire Hydrant	●	■	■	🗄
	Fixed Water Film Foam Extinguishing Device	●	■	🗄	
Power Supply and Distribution System	Power Supply Equipment	●	●	●	🗄
	Power Distribution Equipment	●	●	●	🗄
Central Control Management System	Computer Equipment	●	■	🗄	
	Display Equipment	●	■	🗄	
	Console	●	■	🗄	

Note: ● means mandatory choice, ■ means advised choice, 🗄 means optional choice.

6.3.3 Design of Tunnel Monitoring System

According to the configuration type of tunnel monitoring facilities, combined with the design conditions of tunnel civil engineering, the tunnel monitoring system is designed.

Highway tunnel monitoring system is mainly composed of ventilation and control system, lighting and control system, traffic guidance and control system, fire alarm system, closed circuit television system, cable broadcast system, other systems (such as power supply and distribution system, connecting system) and other subsystems.

Chapter 6 Disaster Prevention and Rescue of Highway Tunnels

1. Ventilation and control systems

The ventilation and control system is mainly composed of axial fan, jet fan, CO/VI detector and wind speed detector. The system mainly monitors the parameters of CO, VI and TW in the tunnel, and controls the fans in the tunnel remotely or manually to meet the ventilation requirements in the tunnel, especially in case of fire or accident (Figure 6-14).

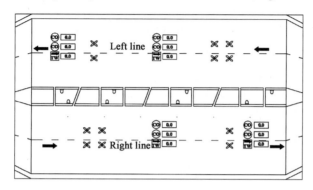

Figure 6-14 Ventilation and control schematics

1) CO/VI tester

The CO/VI detector can automatically detect the CO concentration and fume transmittance in the tunnel. Each equipment shall be arranged in the tunnel section as required, generally arranged at the entrance of the tunnel of 100-200 m, the middle of the tunnel and the exit of the tunnel of 100-200 m, and arranged at 3.5 m above the sidewalk on the right side wall of the driving direction. The spacing between receiving and sending of the detection head shall be 3 m. The period of data acquisition by the CO/VI tester cannot be greater than 60 s.

2) TW tester

The TW detecting device is used for automatically detecting the wind direction and wind speed in the tunnel. The equipment is generally located in the middle of the tunnel and at both ends of the tunnel. The period of data acquisition by the TW tester cannot be greater than 60 s.

3) Jet fan

The jet fan in the tunnel is used to ensure that the fume and CO concentration in the tunnel reach the allowable values and to control the spread of fume in case of fire. The number and type of jet fans are calculated. When the equipment is arranged in the transverse direction of the tunnel, it shall be arranged 15-20 cm outside the construction limit. When the equipment is arranged in the longitudinal direction of the tunnel, it can be arranged centrally at the two end openings for the general long tunnel, and at the two end openings and the middle part of the tunnel for the extra long tunnel.

4) Axial flow fan

The axial flow fan is installed in the wind turbine room (divided into the outside wind turbine room and the inside wind turbine room). Fans shall be selected in combination with operation conditions, ventilation rate, full pressure and performance curve. Axial flow fans with low air pressure and high air volume should be selected. One large fan or several smaller fans in parallel shall be adopted according to the comparison and selection of economic and technical conditions. Blowin type ventilation may not be equipped with diffuser. A diffuser must be installed for suction ventilation. In case of fire in the tunnel, when suction ventilation is adopted, the axial fan shall be able to operate reliably for more than 60 minutes at the ambient temperature of 250 ℃. After recovering to normal temperature, the axial fan can be put into normal operation without overhaul. Axial flow fans do not need to resist high temperature when the wind turbine room outside the tunnel or the wind turbine room inside the tunnel is blown-in ventilation.

2. Lighting and control systems

Tunnel lighting is divided into the following categories: basic lighting, emergency lighting, enhanced lighting (Figure 6-15).

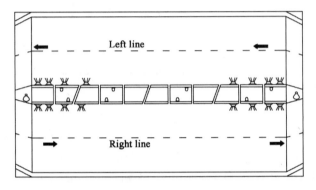

Figure 6-15 Lighting and control schematics

1) Basic lighting

Basic lighting is generally divided into two ways, that is, basic lighting 1, basic lighting 2. The basic lighting 1 is normally open, and it is normally open during the day and at night. Basic lighting 2 is only on during the day.

2) Emergency lighting

Emergency lighting is normally open and always on.

3) Enhanced lighting

Enhanced lighting is divided into entrance section enhanced lighting, exit section enhanced lighting. There are generally four ways to strengthen the lighting at the entrance:

(1) Strengthen the lighting;

(2) Strengthen the lighting;

(3) Strengthen the lighting;

(4) Outlet strengthened lighting generally has 3 ways: enhanced lighting 1, enhanced lighting 2, enhanced lighting 3.

3. Traffic guidance and control system

The traffic guidance and control system is mainly composed of vehicle detector, traffic signal light, lane indicator, variable message board (variable speed limit sign), vehicle crossing sign, etc (Figure 6-16).

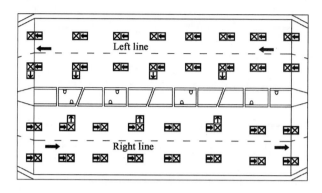

Figure 6-16　Schematic diagram of traffic guidance and control system

1) Vehicle tester

A plurality of vehicle detectors are arranged in the tunnel for detecting traffic flow data passing through the tunnel, and these data are important signs reflecting whether the traffic in the tunnel is normal or not. The vehicle tester can calculate the traffic volume, driving speed, vehicle type, and lane occupancy and so on.

2) Traffic lights

The information machine shows red, green, yellow, and green arrow ends. The red light is the no-go signal, the green light is the pass signal, the yellow flashing light is the attention traveling transition signal, and the red light plus the green arrow is the bypass indication signal.

3) Lane indicator

Lane indicators shall be provided for each lane at the entrance and exit of each tunnel and at the emergency lane. The lane indicator consists of a green arrow and a red X to indicate whether the lane is passable.

4) Variable message board (variable speed limit sign)

To indicate the imminent entry of the vehicle into the tunnel, a variable message board is provided at the entrance to the tunnel. The variable message board can display Chinese characters, letters, numbers and simple graphics. The display attributes are generally more than ten fixed attributes stored in the shelf.

5) Cross channel signs

Across channel sign is set at each channel to guide the vehicles to evacuate safely if an accident takes place in the tunnel.

4. CCTV system

The CCTV system is responsible for monitoring the entire tunnel. Normally used to control traffic conditions. In case of abnormality, it is used to capture the scene image of the emergency scaffold in the tunnel, so that the decision-makers can monitor the scene of the accident remotely, and make the correct rescue and evacuation scaffold scheme (Figure 6-17).

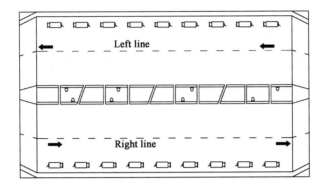

Figure 6-17 Schematic diagram of closed circuit television system

5. Fire alarm system

The fire alarm system consists of centralized fire alarm controller in central control room, fire detector in tunnel, manual alarm button, fume detector in control room, equipment room and substation equipment room, connecting cable, wiring, junction box and necessary accessories. The fire detector and manual alarm button in the tunnel shall be divided into alarm sections, and each section shall be treated as an alarm unit. Segment rack division shall be coordinated with the television magnetic image rack to monitor the range for easy confirmation by the personnel on duty (Figure 6-18).

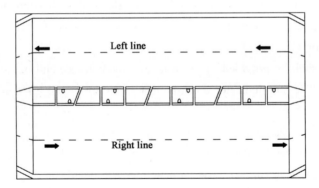

Figure 6-18 Schematic diagram of fire alarm system

6. Cable broadcasting system

Cable broadcast system is a dispatching means for the commanders of the tunnel monitoring center to issue information to the drivers in the tunnel and organize and direct the vehicles and personnel in case of emergency in the tunnel.

7. Other systems

In addition to the above-mentioned subsystems, the tunnel monitoring system also includes: power monitoring system, fire control system, connecting system and so on.

6.3.4 Tunnel Facility Address Code

The address codes of tunnel facilities include: fire area code, ventilation system equipment code, lighting system equipment code, traffic system equipment code, closed circuit television system equipment code, fire alarm system equipment code, cable broadcast system equipment code and other system equipment code.

6.3.5 Number and Position of Tunnel Equipment in Fire

For each fire prevention zone of the tunnel, the number and position of ventilation system equipment, lighting system equipment, traffic system equipment, closed circuit television system equipment, cable broadcast system equipment and other system equipment under fire method shall be determined.

6.3.6 Preparation of Tunnel Disaster Prevention and Rescue Plan

Corresponding disaster prevention and rescue plans shall be formulated for each fire prevention zone in the tunnel. The development process is mainly as follows:

(1) According to the fire prevention zone, determine the number and position of the tunnel equipment under the fire method, and code each equipment.

(2) According to the needs of evacuation and rescue process, determine the execution sequence of each subsystem.

(3) According to the control characteristics of each subsystem, the execution order of each device is determined and sorted by device code. At present, the group sending technology can be used to control the subsystem equipment.

(4) According to the equipment execution order and subsystem execution order, work out the disaster prevention and rescue plan of the fire prevention zone.

6.3.7 Implementation Strategy of Tunnel Disaster Prevention and Rescue Plan

1. Daily tunnel fire management

We will adhere to the principle of "safety first, prevention first" and regularly carry out supervision, inspection, and training exercises. Emergency departments at all levels of each unit shall regularly or irregularly investigate key hazard sources. Intensify supervision and inspection such as investigation and inspection during flood season, thunderstorm season, severe weather such as persistent high temperature, holidays and major activities. Using advanced tunnel electromechanical technology such as tunnel video surveillance, video event detection, fire alarm and so on to do well the prediction and early warning work, timely release early warning information, establish and improve the prevention-oriented daily supervision and inspection mechanism, to avoid and reduce the occurrence of tunnel fire events.

2. Confirmation of tunnel fire

The monitoring center shall make full use of the existing alarm resources such as emergency telephone, tunnel fire automatic alarm, CO/VI automatic alarm and video event detection system as well as the information alarm platform and 110 alarm platform of the expressway operation and management company, accept the system prediction and early warning information and driver and passenger alarm, and carry out linkage alarm.

3. Tunnel fire rescue command system

Once the tunnel fire is confirmed, the tunnel fire rescue command system shall be established immediately, that is, the tunnel emergency command department shall be established (Figure 6-19).

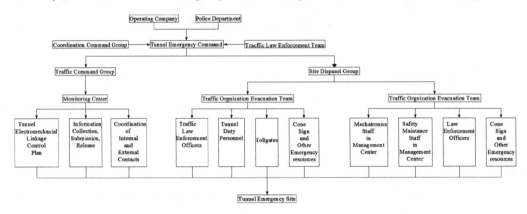

Figure 6-19 Organizational framework of tunnel emergency command

4. Emergency response to tunnel fire

In the event of a tunnel fire, the tunnel emergency command department shall organize emergency resources to make emergency linkage response quickly, be responsible for emergency disposal of tunnel emergencies in the early stage before the arrival of the local government or fire department team, and mobilize the forces of fire control, medical treatment, public security, environmental protection and other social parties to quickly form a joint force for emergency disposal to handle the emergencies.

Upon arrival, the local government or the public security fire department shall hand over the emergency disposal command to the local government or the fire department for on-site general command. The tunnel emergency headquarters shall be responsible for providing traffic support and technical support for rescue work, medical institutions provide medical assistance for emergency disposal, public security organs provide public security for emergency disposal, and environmental protection departments provide environmental protection for emergency disposal, etc. The fire department shall organize on-site rescue and rescue work, and the local government shall organize the emergency disposal work of tunnel emergencies in a unified way.

5. Tunnel fire fighting

In the construction of expressway, fire rescue station can be set up on the road section. After a fire accident, the fire station shall organize the fire fighting force. If no fire station is set up, the local fire department may directly dial 119 to send the nearest fire brigade to rescue and extinguish the fire.

According to the situation of the tunnel under the jurisdiction of the tunnel, fire occurred on the left and right lines of each tunnel, and the rescue route of the external fire brigade was laid out.

6. Remaining Problem Solving after Tunnel fire

Dealing with the remaining problems after tunnel fire includes:
(1) Investigation and evaluation of fire incidents;
(2) Restoration and reconstruction of the tunnel;
(3) Resuming traffic and releasing information;
(4) Public education, training and exercises.

6.4 Disaster Prevention and Rescue System of Tunnel Group

The definition of tunnel group is that there are several continuous tunnels on a road, and there are more than two tunnels between two interchange overpasses, but the tunnels are far away from each other. The ventilation and lighting of the tunnels have no influence on each other, but the traffic flow has certain influence on each other, and the disaster prevention and rescue has great influence. At this time, the control system of each tunnel needs to consider intelligent linkage.

6.4.1 Principles for Linkage Control Plan for Tunnel Groups

1. Section and unit division of the plan

The expressway is divided into plan sections by interchange, and the distance between each adjacent interchange is taken as a plan section. Based on this principle, the expressway is divided into several plan sections which are linked with each other. In each plan section, the section from interchange to interchange, the section from tunnel to tunnel, and the section from tunnel to interchange are taken as plan units along the direction of traffic, bounded by tunnels (single or adjacent tunnels). Based on this principle, the plan section is divided into several interconnected plan units. The plan carries out the overall linkage control when the expressway traffic accidents happen through the monitoring system in the unit. The plan unit is divided into road section unit and tunnel unit (Figure 6-20).

2. Main ideas of linkage control

The main idea of linkage control is that unit linkage control is the foundation and section linkage control is the auxiliary leading idea. Unit linkage control adheres to the control concept of "inside first and then adjacent". Section control adheres to the concept of near-to-far control. The unit linkage control focuses on solving the problems of escape, rescue and traffic organization in and around the accident site, the section linkage control focuses on solving the problem of traffic dispersal.

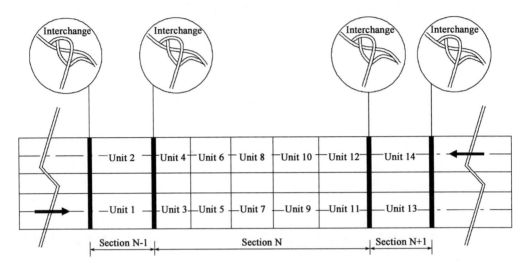

Figure 6-20 Section and unit division of the plan

3. Linkage control strategy

The specific execution order of the plan is: linkage control unit linkage control section linkage control of each equipment control unit monitoring system of the unit monitoring subsystem.

6.4.2 Emergency Response Process of Tunnel Group and External Fire Protection Connection

1. Emergency response process of tunnel groups

The emergency response of tunnel group includes two parts: road section and tunnel. See Section 6.3 for emergency response of tunnel. In case of fire, the rescue organization is as shown in Figure 6-21.

2. External fire connection

Because in the expressway construction, most sections of the road have not set up fire rescue station, so after the fire accident, dial 119 directly, by the local fire department to send the nearest fire brigade to rescue.

6.4.3 Composition of Tunnel Group Monitoring System

The road traffic monitoring system and tunnel traffic monitoring system are usually set along the expressway.

Chapter 6 Disaster Prevention and Rescue of Highway Tunnels

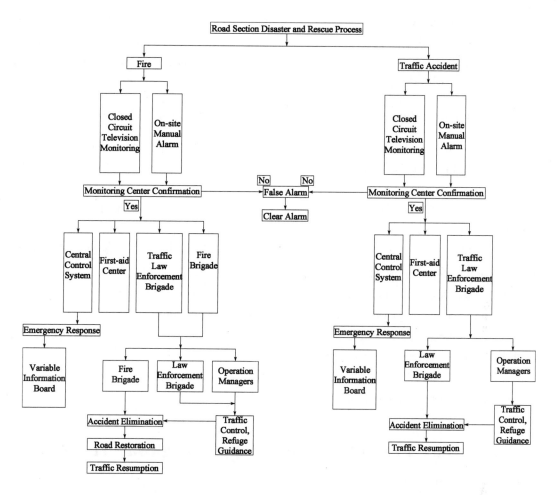

Figure 6-21 Flow Chart of Road Section Disaster Prevention and Rescue

Tunnel traffic monitoring system is mainly composed of ventilation and control system, lighting and control system, traffic guidance and control system, fire alarm system, closed circuit television system and cable broadcast system (Figure 6-22).

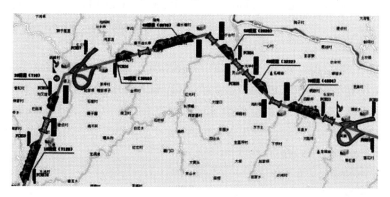

Figure 6-22 Schematic diagram of monitoring system for a highway section

The road traffic monitoring system is mainly composed of variable intelligence, closed circuit television, detectors and other facilities. The road traffic monitoring system is mainly set up in the interworking, service area, parking area, reserved interworking and other important monitoring sections.

6.4.4 Formulation of Fire Emergency Plan for Tunnel Group

The process of formulating a fire emergency plan for tunnel group is mainly as follows:

(1) Divide the highway into sections according to the principle of section division of the plan.

(2) According to the definition of tunnel group, judge whether the relationship between tunnels belongs to tunnel group or not.

(3) The highway is divided into several planning units according to the results of planning section division and tunnel group determination.

(4) Coding the address of the monitoring facilities of the road section unit.

(5) Code that address of the monitoring facility of the tunnel unit.

(6) According to the principle of linkage control plan formulation, the corresponding fire plan shall be formulated for all road section units.

(7) According to the principle of linkage control plan formulation, the corresponding fire plan shall be formulated for all tunnel units.

6.5 Fire Prevention Technology of Highway Tunnel

6.5.1 Classification of Dangerous Goods

Chemicals that have dangerous properties such as explosion, flammability, poison, corrosion and radioactivity and can cause combustion, explosion, personal injury and property loss under certain conditions in the course of transportation, loading and unloading, production, use, storage and safekeeping are collectively referred to as hazardous chemicals. At present, there are more than 2,200 common and widely used species.

In 1986 and 1990, the State Technical Supervision Bureau of China issued the *List of Dangerous Goods* (*GB 6944—86*) and the *List of Dangerous Goods* (*GB 12268—90*), and in 2012, it issued *Classification and code of dangerous goods* (*GB 6499—2012*), which divided dangerous goods into nine categories and specified the names and numbers of dangerous goods. The nine categories of dangerous goods are as follows:

Category 1: explosives;
Category 2: gases;
Category 3: flammable liquids;

Category 4: flammable solids, substances prone to spontaneous combustion, substances that emit flammable gases when saturated with water;

Category 5: oxidizing substances and organic peroxides;

Category 6: toxic and infectious substances;

Category 7: radioactive material;

Category 8: corrosive substances;

Category 9: miscellaneous hazardous substances and articles, including hazardous environmental substances.

The characteristics of fire caused by various dangerous goods are shown in Table 6-4.

Characteristics of fire caused by dangerous goods   Table 6-4

Category Name	Item Name	Basic Characteristics	General Rescue Measures
Explosive	Integral Explosive	When subjected to high heat, friction, impact, vibration or other external influences, it can produce severe chemical reactions, instantly generating a large amount of gas and heat, forming tremendous pressure and causing damage to the surrounding environment.	Leakage: Wet it with water in time and collect it gently after spraying soft material. Fire Hazard: Transfer or isolation; Note: Alkali-acid fire extinguishers or sand are forbidden. Water or other fire extinguishers are available.
	Projectile Explosive		
	Combustion Explosive		
	Normal Explosive		
	Insensitive Explosive		
Compressed & Liquefied Gas	Flammable Gas	When subjected to heat, impact or strong vibration, it can early cause the container to rupture and explode or the gas leakage and cause a fire or poisoning accident.	Leakage: Immediately tighten the valve or immerse the container in cold or lime water; Fire: Sprinkle it with cold water or immerse it in water;
	Non-Flammable Gas		
	Toxic Gas		
Flammable Liquid	Low Flash Point Flammable Liquid	It has high flammability and is characterized by a low flash point and is easily ignited by an open flame or spark at normal temperature. Flammable liquids have high volatility, as well as fluidity and diffusibility which can easily cause burning, explosion and poisoning accidents.	Leakage: Clean up after covering it with dry sand. Fire: Usually not suitable for water;
	Medium Flash Point Flammable Liquid		
	High Flash Point Flammable Liquid		

continued

Category Name	Item Name	Basic Characteristics	General Rescue Measures
Flammable Solid & Pyrophoric Article	Flammable Solid	The flammable solid has a low ignition point, are sensitive to heat, impact and friction and are easily ignited by an external source of ignition. It has a high burning speed and emits toxic and harmful gases or fume solid substances when burned. Pyrophoric article has a low ignition point, is easily oxidized in the air and emit heat and ignites spontaneously. Spontaneous combustion can occur when wet flammable article come into contact with water.	Note: For some metal powders, organometallic compounds, amino compounds, etc., water, foam, carbon dioxide and acid-base fire extinguishing agents are prohibited.
	Pyrophoric Article		
	Wet Flammable Article		
Oxidant & Organic Peroxide	Oxidants	Oxidant can cause intense chemical reactions when it encounters combustible materials and even burns and explodes and at the very least accelerate the combustion reaction. Organic peroxide is flammable, explosive, and easily decomposed and is extremely sensitive to heat, vibration or friction.	Leakage: Rinse it with water after cleaning; Note: When sodium peroxide is on fire, water is forbidden to extinguish it.
	Organic Peroxides		
Toxic & Infectious Article	Toxic Article	After entering the human and animal body, the toxic article can interact with the body tissue, destroys the normal physiological functions of humans and animals, causes temporary or permanent pathological state of the body, and even die. Infectious article contains active microorganisms that cause disease in humans or animals, can cause pathological conditions, and even cause death of humans and animals.	Leakage: Collect it carefully, and clean contaminated vehicles; Fire: Do not use water to extinguish toxic article that reacts dangerously with water; do not use acid-base fire extinguishers for inorganic cyanide.
	Infectious Article		
Radioactive Article		Radioactive article emits a variety of rays that are not perceived by human senses. Excessive exposure to radioactiveparticles may cause damage to the body. In addition, some radioactive particles are extremely chemically toxic.	Leakage: Report to the relevant department immediately.

Chapter 6 Disaster Prevention and Rescue of Highway Tunnels

continued

Category Name	Item Name	Basic Characteristics	General Rescue Measures
Corrosive Article	Acid Corrosive Article	When the corrosive article contacts the human body, it can burn human tissue and cause damage to metal and other items. Irritating gases emitted by corrosive products can cause damage to the eyes and mucous membranes, and cause serious damage to the respiratory tract after inhalation.	Leakage: Sprinkle it with dry sand, sweep clean and rinse with water;
	Alkaline Corrosive Article		
	Other Corrosive Article		
Other Flammable Article		Such as wood, cloth, paper, animal hair, carbon black, coal powder, etc.	

6.5.2 Management Methods for Safe Transportation of Vehicles Carrying Dangerous Goods through Tunnels

1. General principles

Article 1 The measures are formulated in accordance with relevant laws and regulations and in combination with the actual situation of highway tunnels, which help to strengthen fire control supervision and management, prevent and reduce fire, protect social property and personal safety, and ensure the safe passage of vehicles through the tunnels.

Article 2 The measures are applicable to extra-long highway tunnels.

Article 3 In addition to these measures, all vehicles passing through the tunnel shall abide by the relevant standards and norms currently in force at the State and the Ministry of Transport

Article 4 In order to strengthen the fire safety management of tunnels, the fire control policy of "prevention first, combination of prevention and elimination" and the idea of "prevention before danger" shall be implemented to ensure the safe passage of vehicles carrying inflammable and explosive materials [see *Classification and code of dangerous goods* (*GB 6499—2012*) for details] through the tunnels. Mainly includes three aspects: the human, the equipment, the management.

Article 5 It is the common responsibility and obligation of all tunnel users to maintain the tunnel public fire safety, protect the fire-fighting facilities, and prevent and put out fires.

Article 6 Matters not covered by these measures shall be implemented in accordance with the relevant provisions of the Ministry of Connecting.

Article 7 The power of interpretation of these Measures shall be vested in the administrative department of extra-long highway tunnels.

2. Basic systems

Article 8 Post Responsibility System: The administrative department of extra-long highway tunnels shall establish the post responsibility system for the safety inspectors who prevent the fire of inflammable and explosive materials, specify the work contents, rights and responsibilities, and carry out the work with certificates.

Article 9 Work record system: In addition to recording the safety status of vehicles with flammable and explosive materials allowed or restricted to pass through, the security inspectors shall also record the alarm equipment, fire-fighting equipment and monitoring system in the tunnel, summarize them in time, find out the problems as soon as possible and reflect the problems.

Article 10 Forecast system: When the security inspectors discover the hidden fire hazards, they shall promptly react to the tunnel management department and immediately organize relevant personnel to clear the hidden fire hazards.

Article 11 Reward and Punishment System: Meritorious service shall be rewarded to those who are outstanding in their professional level and who have been conscientiously responsible for their work or who have discovered and eliminated the hidden danger of major fires. Those who are irresponsible for their work or fail to meet their professional standards or lose their jobs shall be given disciplinary actions such as warning, fine, demerit recording, retainer inspection, dismissal and so on, depending on the seriousness of the case, and shall be investigated for criminal responsibility.

Article 12 System of summary analysis and analysis: Safety reports shall be carefully filled in and compiled every month, the safety situation of extra-long highway tunnels in this month shall be summarized and analyzed, and the hidden fire hazards discovered during the inspection shall be registered and put on record one by one.

Article 13 Regular drilling system: The administrative department of extra-long highway tunnel shall organize regular fire rescue drills by security inspectors, drivers and local fire control departments to raise people's awareness of the importance of tunnel fire prevention, familiarize themselves with the use of fire-fighting equipment, understand the procedures of disaster prevention plans, and improve their fire rescue capability in emergency situations.

Article 14 Regular maintenance system: Establish and improve the large-scale maintenance and verification of the equipment in the tunnel every quarter, ensure the reliability of the signal and connecting power supply equipment, ensure the smooth flow of the connecting, signal and power supply lines, strengthen the maintenance and repair of the driving safety facilities, and improve the safety guarantee capability.

Article 15 Regular training system: With the overall improvement of the quality of the staff and workers as the fundamental measure, great efforts shall be made to train all the staff and workers, especially the key posts and on-the-job personnel. Professional training for key positions and incumbents should be conducted once every two years in order to update knowledge and continuously improve the professional quality of the staff.

Article 16 Inter-departmental connecting system: The administrative department of extra-long highway tunnels shall strengthen working contact with the local fire control department, jointly study and solve the problem of preventing and rescuing vehicles of inflammable and explosive materials from passing through the extra-long highway tunnels, jointly establish a strict fire safety system, and carry out regular tests on the formulated fire rescue procedures.

Article 17 Inter-departmentalconnecting system: The administrative department of extra-long highway tunnel shall strengthen working contact with the local fire control department, jointly study and solve the problem of preventing and rescuing vehicles of inflammable and explosive materials from passing through the extra-long highway tunnel, jointly establish a strict fire safety system, and carry out regular tests on the formulated fire rescue procedures.

Article 18 Firefighting organizations: An effective firefighting organization shall be established in the extra-long highway tunnel according to the fire characteristics of inflammable and explosive materials, and sufficient firefighting equipment shall be equipped, and the equipment shall be inspected and replaced regularly to maintain the normal operation of the equipment. At the same time, corresponding fire rescue plans should be made according to the possible fire accidents.

3. Safety inspection of vehicles carrying dangerous goods

Article 19 The administrative department of extra-long highway tunnels shall set up checkpoints for inflammable and explosive articles at both ends of the tunnels. Checkpoints shall include two separate sections of the inspection area and the area where temporary parking has been inspected, with unobstructed road connections to tunnel entrances and the road network. Checkpoints can generally be considered in conjunction with tunnel tollbooths to determine their location and size.

Article 20 Adequate dangerous goods inspection equipment shall be set up at the checkpoints, and periodic inspection and maintenance shall be carried out. Safety inspectors shall be proficient in the testing equipment and methods for all kinds of dangerous goods.

Article 21 Checkpoints shall be equipped with perfect safety facilities and disaster prevention measures for lighting, alarming, firefighting, explosion prevention, lightning protection and elimination of static electricity.

Article 22 Dangerous goods vehicles shall be temporarily parked at checkpoints, stored in different areas and equipped with corresponding disaster prevention equipment.

Article 23 Functions of dangerous goods inspection: (1) To detect hidden dangers of dangerous goods vehicles, and to remedy and strengthen them; (2) Temporary parking of vehicles with inflammable and explosive materials and organization of safe passage of vehicles through the tunnel.

Article 24 Safety Standards for Vehicles Passing Through Tunnels with Flammable and Explosive Materials: (1) The traffic monitoring, ventilation, lighting, alarming and fire control systems of the tunnels are in good condition; (2) The loading and packaging of dangerous goods vehicles meet the standards, and the safety measures are appropriate; (3) Appropriate passing time and surrounding environment.

Article 25 Personnel shall strictly abide by the safety standards formulated by the tunnel management department, According to the safety situation of vehicles loaded with flammable and explosive substances, the risks of vehicles loaded with flammable and explosive substances passing through tunnels and bypassing schemes are comprehensively weighed from the aspects of economy, safety and society, and dealt with according to three situations: (1) vehicles loaded with Class 9 substances are allowed to pass through the tunnels; (2) Vehicles loaded with substances of Categories I, II, III, IV, V, VI and VIII shall be strictly restricted in their passage according to the safety measures adopted; (3) Vehicles carrying Category I and Category VII substances are prohibited from passing through the tunnel. Vehicles shall be required to leave the checkpoint and drive in a detour immediately after the determination of the test.

Article 26 The tunnel management department shall have the right to require the vehicles transporting dangerous goods to provide such parameters as scientific name, alias, physical and chemical characteristics, main components, packaging method, transportation precautions and danger of the goods, as well as the safety and security measures taken by the vehicles.

Article 27 The tunnel management department shall conduct short-term training for drivers and freight escorts of temporarily parked vehicles to explain the transportation instructions in the tunnel, the use of disaster prevention equipment and the fire rescue methods in case of emergency.

4. Fire Control of dangerous goods vehicles passing through tunnels

Article 28 The maintenance and repair of major equipment shall be carried out in accordance with the principle of hierarchical responsibility and centralized management by departments. Clear responsibilities, and establish the corresponding classification system.

Article 29 The duties of the administrative department are to lead and organize the work of fire prevention and safety management for vehicles with inflammable and explosive materials passing through the tunnel, and to organize the security inspectors to study the relevant rules and regulations, so as to improve their professional quality. Summarize experience, carry out evaluation, implement rewards and punishments, and improve the management level in an all-round way.

Article 30 During fire-prone seasons and major festivals, the administrative departments shall organize special forces to carry out propaganda and education on fire control and safety inspection, and do a good job in fire control.

Article 31 To strengthen unified leadership, implement comprehensive management and bring work safety into the orbit of modern management.

Article 32 The safety supervision office of the tunnel management department shall regularly inspect and evaluate the safety work of the tunnel, and propose new measures and methods according to the operation of the tunnel.

Article 33 Full-time security inspectors in extra-long highway tunnels shall be responsible for the safety inspection in the tunnels and before vehicles enter the tunnels, so as to detect hidden fire hazards as soon as possible. The security personnel shall undergo the professional training of tunnel fire control before taking up their posts. After passing the training, they shall take up their posts with certificates and maintain the relative stability of the personnel. Security personnel should be trained professionally once every two years to update their knowledge.

5. Disaster prevention measures for dangerous goods vehicles passing through tunnels

Article 34 With the development of industry, the types and volumes of dangerous goods are increasing year by year, making it difficult to identify dangerous goods, determine their safety levels and take appropriate safety measures. Therefore, the tunnel management department should establish the dangerous goods safety information management system, provide information query, safety decision-making and emergency rescue plan and other services, in order to improve the modern management level of the tunnel.

Article 35 Vehicles of inflammable and explosive substances allowed or restricted to pass through the tunnel may only be guided through the tunnel by the safety vehicles of the tunnel management department within a specified period of time. The transit time shall be selected when there are few vehicles according to the traffic volume.

Article 36 When the vehicles with inflammable and explosive substances pass through the tunnel, the tunnel security personnel and the central control room shall monitor the running status of the vehicles in real time and find out the problems in time. Alarm, fire control, ventilation and other systems in the tunnel shall be prepared for emergency.

Article 37 Structural measures: (1) Guardrails shall be installed on both sides of the entrance of extra-long highway tunnel, and the entrance shall be provided with good lighting equipment and obvious signs and protective facilities; (2) sufficient transverse slope should be set in the tunnel to facilitate drainage and keep the road surface dry; (3) Oil collecting trough and flammable liquid channeling trough should be set in the tunnel, so that the flammable liquid leaked out is limited in a small range, the evaporation of flammable liquid is reduced, and the explosion risk is reduced.

Article 38 Measures for traffic control: (1) Separation of passenger and freight lanes. The key point of security inspection is that after the trucks and vans are separated from each other, the operation of passenger cars will be smoother. (2) Smoking and carrying open flames are prohibited in the tunnel; (3) No parking or overtaking is allowed in the tunnel. The truck spacing shall be guaranteed to be more than 100 m; (4) Drunk driving is prohibited.

Article 39 A special wireless connecting network system shall be established in the area of extra-long highway tunnels to ensure good connecting between drivers, firefighters and the central control room in case of safety monitoring or emergency. Drivers and security personnel in the tunnel can receive guidance from the central control room at any time. At the same time, the central control room can also be contacted by wireless interphone and emergency telephone set in the tunnel.

Article 40 Emergency energy supply shall be provided in the tunnel. Emergency energy supply can be provided by backup power supply or generator, which should provide sufficient energy for tunnel lighting, ventilation, traffic control, connecting and other important systems.

Article 41 Extinguishment of flammable and explosive dangerous goods in emergency: (1) According to the nature of the goods, the correct methods shall be adopted to organize the extinguishment. For flammable liquids and aluminum-iron fluxes, no water shall be used to put them out; No sand shall be used to cover the explosives; Rescuers shall take anti-poisoning and anti-corrosion measures for dangerous drugs, corrosive articles or articles that produce toxic gases when burned. (2) Dangerous goods vehicles that are subject to fire must be completely put out, especially cotton, hemp, wool and other articles, and the latent fire must be thoroughly inspected and eliminated. (3) Adopt isolation measures to prevent the disasters caused by explosive materials from spreading.

 **Exercise**

6.1 Brief description of tunnel area division and fire stage division in case of fire.

6.2 What are the distributions of the longitudinal and transverse temperatures of the tunnel in case of fire?

6.3 What are the reasons for the countercurrent of flue gas and the underlayer of flue gas?

6.4 Describe briefly the ventilation and fume control standard for fire scenarios in double tunnels.

6.5 Briefly describe the fume control standard for single tunnel with ventilation shaft combined fire scenarios.

6.6 What are the monitoring systems for tunnels?

6.7 Brief the formulation process of disaster prevention and rescue plan for single tunnel.

6.8 Brief description of section and unit division of tunnel group plan.

6.9 Brief description of the preparation process of disaster prevention and rescue plan for tunnel group.

6.10 What are the categories of dangerous goods? What's the difference?

 Key vocabulary:

chimney effect 烟囱效应	closed circuit television system 闭路电视系统
fire attenuation stage 火灾衰减阶段	fire development stage 火灾发展阶段
fire stabilization stage 火灾稳定阶段	Fire wind pressure 火风压
emergency response 紧急响应	linkage control 联动控制
oxygen-rich combustion 富氧燃烧	throttling effect 节流效应
traffic wind pressure 汽车通风力	tunnel groups 隧道群

Chapter 7　Railway Tunnel Disaster Prevention and Rescue

[Important and Difficult Contents of this Chapter]
(1) The development course of disaster prevention and rescue in railway tunnel.
(2) Composition of rescue and evacuation facilities.
(3) Ventilation and disaster prevention scheme design of railway tunnel.
(4) Arrangement of ventilation equipment in railway tunnel.
(5) Design of wind speed and air volume for disaster prevention and rescue of railway tunnel.

7.1　Development of Disaster Prevention and Rescue in Railway Tunnels

With the rapid development of railway construction in China, tunnel has been widely used because of its advantages of shortening the mileage of railway and reducing environmental damage. However, tunnel fires occur from time to time, which seriously threaten the safety of passengers' lives and property, and even cause huge social impact and economic losses. Because of the enclosure of tunnel environment, the condition of fume exhausting and heat dissipation is poor, and the temperature is high. The high concentration of toxic fume will be produced quickly, which makes evacuation difficult, disaster relief difficult, and the serious damage degree. At present, when a railway tunnel fire occurs, the train is dragged out of the tunnel as far as possible for firefighting and rescue. With the development of railway tunnels, trains may be supply to stop without power before being towed out of the tunnels. Therefore, in the background of the tunnel construction gradually in-depth, to ensure the safe operation of the tunnel has become the focus of public concern, and also become the focus of ventilation and disaster prevention design. In order to standardize the design of disaster prevention ventilation for domestic railway tunnels, the State Railway Administration issued *Code for Design on Rescue Engineering for Disaster Prevention and Evacuation of Railway Tunnel* (*TB 10020—2017*) on May 1, 2017.

Railway tunnel is a long and narrow structure, once the train fire in the tunnel, it will produce a large number of toxic fume and hot fume, which will pose a great threat to human life and the tunnel, the consequences are unimaginable. There have been more than 30 serious train fires at home and abroad. For example, in 1972, the North Land Tunnel in Japan was hit by a passenger train fire caused by leakage of electricity from electrical equipment, resulting in more than 700 casualties. Since 1976, there have been seven major tunnel train fires in China, with a total suspension time of 2500 hours, more than 300 casualties (115 deaths) and direct economic losses of more than 30 million CNY, while indirect economic losses cannot be estimated.

At home and abroad, the research on railway tunnel fire can be divided into three aspects: solid-size fire test, fire model test and numerical simulation.

Countries all over the world have invested a lot of manpower, material resources and financial resources in the field test and research of railway tunnel fire. In 1965, Switzerland conducted a large-scale fire test in the abandoned Offenegg railway tunnel near Wesson. The aim is to measure the thickness of the fume layer inside and outside the tunnel entrance. In 1973, a series of fire tests were carried out on open-air railway lines in Japan. The test results show that the train is safe to run for 15 minutes after the fire. In 1974—1975, Austria carried out fire tests in the abandoned Zwenberg tunnel. It was found that different ventilation methods had great influence on fuel combustion, fume direction and fume temperature. From 1990 to 1993, nine countries in Western Europe jointly conducted large-scale fire tests in tunnels in Germany, Norway and Finland to test the temperature, heat conduction, fume flow rate, fume concentration and their effects on visibility in the whole tunnel.

In recent years, with the development of computer technology and computational mathematics, the use of computer simulation of tunnel fire has been favored by more and more scholars. Compared with the experimental study, the computer simulation of tunnel fire has the advantages of arbitrariness of parameter setting and reproducibility of prediction results. At present, the computational fluid dynamics (CFD) method and the fire numerical calculation program are mainly used to simulate the flow of high temperature fume and the distribution of toxic gases in various tunnel fires. In 1996, Woodburn and Britter used CFD method to simulate the temperature and flue gas near the tunnel fire source, that is, in the downwind direction of the fire source, and focused on the influencing factors of the numerical simulation results.

7.2 Rescue and Evacuation Facilities

7.2.1 Cross Channel and Parallel Adit

Cross channels shall be provided between the two parallel tunnels, and the distance between the cross channels shall not be greater than 500 m. The cross channel design shall comply with the following provisions:

(1) The short-lived dimension should not be less than 4.0 m × 3.5 m (width × height).

(2) The cross channel shall be provided with a protective door that is easy to open.

(3) The net width of the guard door shall not be less than 1.5 m and the net height of the guard door shall not be less than 2.0 m.

(4) Longitudinal gradient shall not be greater than 1%, and the opening range of the protective door shall be flat slope.

When the construction organization of a single tunnel requires the installation of parallel pilot holes, the parallel pilot holes shall be used as rescue evacuation tunnels. The cross-sectional dimension of parallel pilot pit shall not be less than 4.0 m × 5.0 m (width × height).

The ground of the transverse gallery and the parallel adit shall be flat, stable and free of water accumulation.

Installation of disaster prevention ventilation, emergency lighting, emergency connecting and other facilities shall be provided for parallel adit and cross channels.

7.2.2 Emergency Exit

Emergency outlet can be divided into vertical shaft type emergency outlet, inclined shaft type emergency outlet, transverse gallery type emergency outlet.

The emergency exit shall be designed in accordance with the following provisions:

(1) Vertical shaft emergency exit: The vertical height should be less than 30 m, and the total staircase width should be less than 1.8 m.

(2) Inclined shaft emergency outlet: when the slope is not greater than 12%, the transverse length should not be greater than 500 m, and when the slope is not greater than 40%, the transverse length should not be greater than 150 m.

(3) Cross channel emergency outlet: the length should not be greater than 1000 m.

(4) The connection between the emergency exit and the main tunnel shall be provided with a protective door which is easy to open, and the width shall not be less than 1.5 m, and the height shall not be less than 2.0 m.

(5) The cross-sectional dimensions of inclined shaft type and cross channel type emergency outlet shall not be less than 3.0 m × 2.2 m (width × height), and the dimensions of vertical shaft type emergency outlet shall be determined according to the staircase arrangement.

When setting emergency exit for a single tunnel, the transverse gallery shall be preferentially selected, and the shaft or inclined shaft that meets the requirements may also be selected.

The ground in the emergency exit gallery shall be flat, stable and free of water. The emergency exit shall be equipped with disaster prevention ventilation, emergency lighting, emergency connecting and other facilities.

7.2.3 Shelter

The shelter shall be designed in accordance with the following provisions:

(1) The cross-sectional dimension of the access adit for the shelter should not be less than 4.0

m × 5.0 m (width × height).

(2) Protective doors shall be set at the junction of the access adit for the shelter and the main tunnel. The clearance width of the protective doors shall not be less than 1.5 m and the height shall not be less than 2.0 m.

(3) The net space in the shelter shall be determined according to the nature of the passenger train and the regional characteristics, etc. The number of people to retreat shall be determined according to the percentage of the number of passengers, and the area per capita shall be considered as 0.5 m^2/person.

(4) The slope of the tunnel bottom of the access adit for shelter and the space for retreat shall not be greater than 3%, and the opening range of the protective door shall be flat slope.

The ground in the shelter and tunnel shall be flat, stable and free of water. Disaster prevention ventilation, emergency lighting, emergency connecting and other facilities shall be provided in the shelter and tunnel.

7.2.4 Emergency Rescue Station

The cross channel design shall include the following:

(1) Determine the length, platform width and height of the emergency rescue station.

(2) Calculate the spacing (density) of the cross channel, the net width and height of the cross channel door.

(3) Calculate the waiting area according to the rescue method.

(4) Determine disaster prevention ventilation, emergency lighting, emergency connecting, fire control and other facilities and equipment.

The length of emergency rescue station is generally 550-600 m, which is composed of passenger trains and a certain amount of spare capacity. The tunnel length of high-speed railway and passenger dedicated line should be 450-500 m for moving train sets only. The evacuation platform shall be set up in the emergency rescue station. The width of the platform shall be 2.3 m. The dimension of the platform surface higher than the track surface shall not be less than 30 cm, and shall not intrude into the building boundary. The distance between the evacuation cross channels of the emergency rescue station shall not be greater than 60 m, and the dimension of the cross channels in the emergency rescue station shall not be less than 4.5 m × 4.0 m (width × height). Protective doors shall be set at both ends of the cross channel in the emergency rescue station. The total width of the clearance of the protective doors shall not be less than 3.4 m, the clearance height shall not be less than 2.0 m, the longitudinal gradient of the cross channel shall not be greater than 1%, and the opening range of the protective doors shall be flat slope.

When the rescue train is used to rescue out of the tunnel, the waiting area for personnel in the emergency rescue station shall be 0.5 m^2/person. The ground in the emergency rescue station shall be flat, stable and free of water. The emergency rescue station shall be equipped with disaster prevention ventilation, emergency lighting, emergency connecting and fire control facilities.

Chapter 7 Railway Tunnel Disaster Prevention and Rescue

7.3 Ventilation Design for Tunnel Disaster Prevention

Code for Design on Rescue Engineering for Disaster Prevention and Evacuation of Railway Tunnel (*TB 10020—2017*) stipulates that emergency rescue stations shall be set up for tunnels or groups of tunnels with a length of 20 km or more, and the distance between emergency rescue stations shall not be greater than 20 km. The emergency rescue station shall be equipped with conditions for the rapid evacuation of personnel to a safe area and the ability to rescue themselves or to reach out of the hole through self-rescue. A single tunnel with a length of 10 km or more shall have at least one emergency exit or shelter in the tunnel section. For a single tunnel with a length of 5-10 km, an emergency exit or shelter shall be set up in the tunnel section. For a single tunnel with a length of 3-5 km, an emergency exit can be set in the tunnel section in combination with the construction access adit. According to the requirements of *Code for Design on Rescue Engineering for Disaster Prevention and Evacuation of Railway Tunnel* (*TB 10020—2017*), appropriate disaster prevention ventilation design schemes are adopted for different tunnel conditions.

7.3.1 Ventilation for Disaster Prevention of Tunnels with Rescue Stations

The disaster prevention ventilation system of long tunnels can be combined with the longitudinal ventilation of the main tunnel and the transverse ventilation of the rescue station. A bidirectional reversible jet fan is arranged in the main line tunnel in niche arrangement (see Figure 7-1); A transverse ventilation duct is arrange in that rescue station, and the transverse ventilation duct is connected with the ventilation incline shaft. A rescue station is arranged in the long tunnel. The rescue station divides the tunnel into different disaster prevention areas. The inclined shaft at the rescue station and the jet fans at both ends of the tunnel are used to supply air or exhaust fume for the rescue station. Ventilation must be maintained

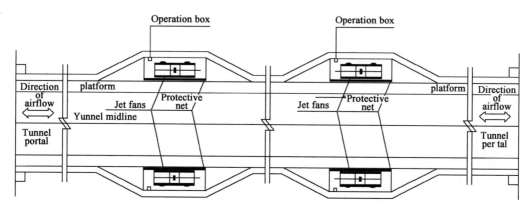

Figure 7-1 Arrangement of longitudinal ventilation jet fan in main line

The cross channel of the rescue station is under positive pressure to prevent the flue gas from flowing into the gallery. The ventilation system supplies air from the safety tunnel, and the fresh air flows to the direction of the accident tunnel through the cross channel, restrains the fume from entering the cross channel and the tunnel without accident, guides the evacuees to enter the cross channel and the safety tunnel in the face of the fresh air, and ensures the fresh air required by the personnel in the area to be avoided. Different disaster prevention ventilations are adopted according to the different areas in the tunnel where the accidents occur. When a train runs in a tunnel and needs to stop at a rescue station for rescue, a reversible jet fan is arranged in all the connecting cross channels within the rescue station to obtain fresh air from the safety tunnel to supply air in the accident duct. Figure 7-2 and Figure 7-3 are ventilation for disaster prevention.

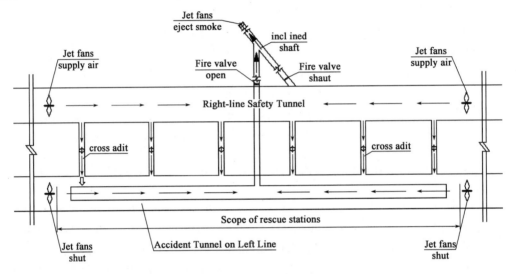

Figure 7-2 Disaster prevention ventilation schematics of rescue station when accident in left-line tunnel

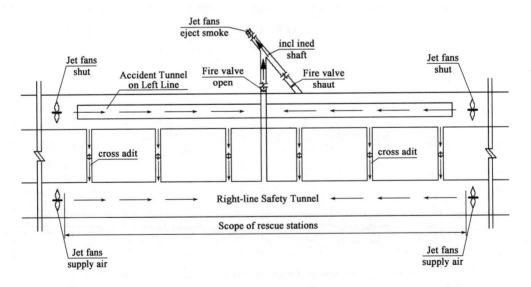

Figure 7-3 Disaster prevention ventilation schematics of rescue station when accident in right-line tunnel

7.3.2 Disaster Prevention Ventilation of Tunnels with Emergency Exits or Shelters

For tunnels requiring emergency exits or shelters, construction inclined shafts are usually used as emergency exits or shelters [see Figure 7-4a)]; Others widen the cross-section of the inclined shaft as a temporary shelter [see Figure 7-4b)]; Or as an emergency shelter in an inclined shaft near the tunnel [see Figure 7-4c)]. A fan is arranged in the inclined shaft, and fresh air is pressurized in the emergency outlet in case of accident, so as to maintain positive pressure in the emergency outlet and the shelter area, inhibit fume from entering the shelter, and ensure the safety of evacuees. With regard to the programme to establish parallel caverns as safe havens, Because of the large depth of the parallel pilot pit, the air flow is not circulating and the air quality is poor, which cannot meet the requirements of the fresh air provided to the asylum personnel by the standard. Therefore, it is necessary to set the blower at the intersection of the parallel pilot pit and the inclined shaft, and send the fresh air sent by the blower into the inclined shaft to the shelter area through the way of "secondary relay". When there is a fire in the tunnel and evacuation is required by emergency exit, turn on the pressurized blower in the emergency exit, and turn on the blower serving the shelter when there is a parallel adit as the shelter. When the pressure difference between the emergency exit and the tunnel reaches 40-50 Pa under the action of the pressurized blower, the residual pressure valve at the protection door is automatically opened to relieve the pressure, so as to ensure the smooth opening of the protection door and facilitate the evacuation of personnel.

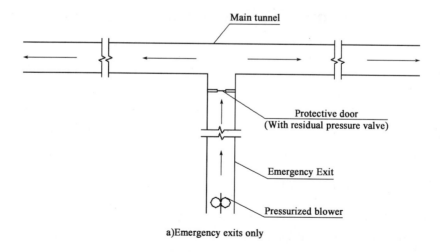

a) Emergency exits only

Figure 7-4

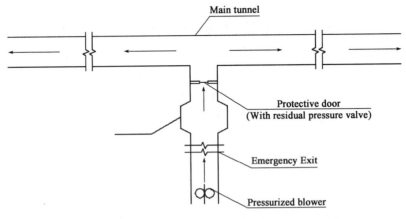

b)Establishment of shelters (partial widening of emergency exits)

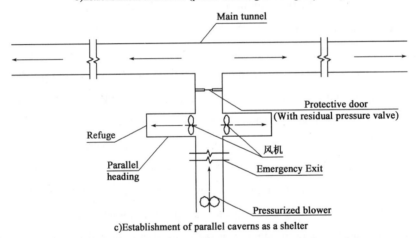

c)Establishment of parallel caverns as a shelter

Figure 7-4　Disaster prevention ventilation with emergency outlets or shelters

7.4　Ventilation Equipment Arrangement

Most of the jet fans in the tunnel are arranged on both sides of the tunnel in niche style. When the number of jet fans is large, they should be arranged in groups with a certain distance. This method needs to widen the tunnel section locally and increase the tunnel investment. Lifting, direct stacking or niche placement can be adopted when the pressurizing fan is installed in the inclined shaft, depending on the tunnel section and whether the gallery is affected.

7.4.1　Niche Arrangement

Fan equipment with niche arrangement needs to widen the tunnel section locally, and the widening amount of the section is determined according to the type and size of the fan, outlet wind speed, installation requirements and other factors. Niches-type arrangement of jet fans is shown in Figure. 7-5.

Chapter 7 Railway Tunnel Disaster Prevention and Rescue

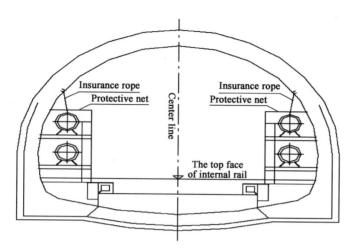

Figure 7-5 Schematic diagram of jet fan for niche arrangement

As far as the fan support is concerned, it is usually made of steel structure, which is uniformly configured by the fan manufacturer with the equipment or separately purchased and manufactured by the equipment installation unit. The steel bracket is made according to the actual size of the fan and the space size of the installation position in the later stage of installation, which is convenient and flexible. However, the steel bracket is easy to be corroded due to the long-term humid environment in the tunnel, which affects the service life, safe operation of the fan and heavy workload of operation and maintenance. Concrete supports do not corrode, so it is suggested to use concrete supports, but the construction of concrete supports is not as flexible and convenient as steel supports.

The construction of concrete bracket can be divided into two cases: One is to make the fan bracket well together when constructing the tunnel, but because the equipment is not tendered, there are uncertain factors to make the bracket ahead of time. The other is to reserve reinforcements at the place where the fan is required to be installed, and the installation unit shall separately construct and manufacture the fan at the later stage. However, the equipment installation unit is generally not good at the construction of concrete structure and cannot guarantee the construction quality, and the construction unit is reluctant to separately purchase less construction materials. Therefore, it is suggested to adopt the method of containment design, and the construction unit should do it well at one time during tunnel construction, leaving enough space for equipment transportation, installation and overhaul.

7.4.2 Hoisting Arrangement

The inclined shaft usually passes less, and there are not too many hidden dangers in hoisting the fan. Therefore, the usual way is to install the fan in the inclined shaft with the suspended ceiling, which can save the professional investment of the tunnel. However, because of the large height of inclined shaft (generally more than 5 m), it is difficult to install, maintain, overhaul and replace the suspended ceiling. Although the initial investment is saved, the operation cost is higher in the later period. Therefore, it is not recommended to install the ceiling, but to adopt niche-type installation. Hoisting arrangement of jet fans is shown in Figure. 7-6.

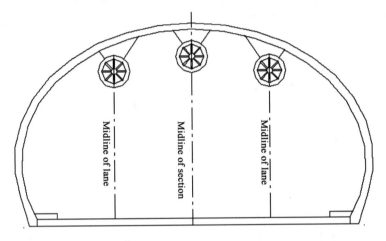

Figure 7-6　Schematic diagram of jet fan for hoisting arrangement

7.5　Design of Wind Speed and Volume for Disaster Prevention and Rescue

7.5.1　Critical Wind Speed in Tunnel Fire

The critical wind speed for disaster prevention ventilation in domestic tunnels is not less than 2 m/s. The calculation formula proposed by Kennedy is usually used to calculate the critical wind speed in foreign tunnels:

$$V_c = K_g \cdot K \frac{(gHQ)^{\frac{1}{3}}}{\rho_0 c_p A_r T_f} \tag{7-1}$$

$$T_f = \frac{Q}{\rho_0 c_p A_r V_c} + T_0 \tag{7-2}$$

Where: V_c——the critical wind speed in the longitudinal ventilation tunnel, m/s;

　　　g——the gravitational acceleration, m/s^2;

　　　H——that tunnel height in the fire area, m;

　　　Q——the heat release rate of fire, kW;

　　　ρ_0——the air density, kg/m^3;

　　　c_p——the specific heat of air at constant pressure, kJ/(kg · K);

　　　A_r——the cross-section area of the tunnel, m^2;

　　　T_f——the temperature of hot air, K;

T_0——ambient air temperature, K;

K——dimensionless parameter, take 0.61;

K_g——the slope correction coefficient, $K_g = 1.0$ for $i \geq 0$; $K_g = 1 + 0.0374 i^{0.8}$ for $i < 0$.

In case of fire in the tunnel with longitudinal fume exhaust, the air volume required to maintain the critical wind speed is $L = A_r \cdot V_c$.

7.5.2 Calculation of Wind Speed Distribution in Fan Niche Arrangement

The jet fan arranged in niche is close to the evacuation platform on the side of the tunnel wall. When the fan is running, the wind speed blowing to the evacuation platform should be within the range that pedestrians can bear to ensure the safety of evacuees. The installation position of the fan shall be determined after wind speed accounting. Because the niche-type fan is close to the tunnel wall, the irregular tunnel wall will inevitably affect the flow movement of the fan outlet, and the characteristics of the flow movement are complex. The calculation results can be modified by using the wind speed calculation formula of circular cross-section jet movement for reference:

$$V_1 = K \cdot \frac{0.095}{\frac{as}{d_0} + 0.147} \cdot V_0 \qquad (7-3)$$

Where: V_1——the calculated wind speed, m/s;

a——the turbulence coefficient, dimensionless;

s——the distance from the outlet of the fan to the calculated point, m;

d_0——the outlet diameter of the fan, m;

V_0——the average wind speed at the outlet of the fan, m/s;

K——a modified parameter, dimensionless, which is related to the influencing factors of fan jet space.

The wind speed distribution in fan niche arrangement can also be obtained by numerical simulation. However, due to the influence of many factors, it is difficult to reflect the method one by one, and the setting of some boundary conditions is quite different from the actual situation. Therefore, the calculation results are only approximate, and it is necessary to analyze the calculation results according to the actual situation, and make appropriate corrections to the results.

7.5.3 Design of Wind Speed and Volume at Rescue Station, Emergency Evacuation Outlet and Shelter

(1) The wind speed at the guard door of the cross channel of the rescue station shall not be less than 2 m/s. If the space to be avoided is set, the fresh air volume of the space to be avoided shall meet the requirements of 10 m³/(person·h).

(2) During the operation of the fan in the emergency evacuation port, the wind speed distribution can be calculated by Formula (7-3) or determined by the method of simulation, so as

to determine the selection of the fan and the installation position of the fan.

(3) The wind speed at the guard door in the junction of the emergency evacuation entrance and the tunnel shall not be less than 2 m/s. If there is a shelter, the fresh air volume of the shelter shall meet the requirements of 10 m^3/(person · h).

7.6 Cooperation of Ventilation with Other Specialties

The successful completion of tunnel ventilation design requires close cooperation with relevant specialties, including collecting and sorting the data for some specialties and providing relevant information and materials for other specialties.

(1) Cooperation with tunnel specialty. Tunnel fans with niche arrangement need tunnel professionals to widen the tunnel section according to the arrangement requirements of fans, and the longitudinal length of widening section should meet the ventilation requirements. Fan installation place shall be reserved with concrete support and hoisting fan hooks. After installation, the fan shall not intrude into the pedestrian platform on the side of the tunnel wall. The fans to be hoisted shall be steel plates specially reserved for installation of fans in the tunnel. Ventilation specialty should give maximum load to tunnel specialty.

(2) Cooperation with economic adjustment specialty. Fresh air volume shall be calculated according to the number of persons to be sheltered who need to be provided with shelter by the transferred specialty.

(3) Cooperation with power distribution specialty. Ventilation professionals and power distribution professionals connect to determine the best location of the fan set, in order to meet the ventilation requirements, but also to facilitate power supply to reduce investment.

(4) Cooperation with architectural specialty. Set up ventilation equipment management room and spare parts warehouse on site or in the whole line operation management center according to operation requirements. Raise housing allocation requirements for building professionals according to ventilation requirements.

(5) Cooperation with monitoring specialty. *Code for Design on Rescue Engineering for Disaster Prevention and Evacuation of Railway Tunnel* (*TB 10020—2017*) requires that a monitoring system for disaster prevention and rescue equipment should be designed for tunnels equipped with disaster prevention and ventilation, and the system should have remote control functions. As an integral part of tunnel disaster prevention and rescue, disaster prevention and ventilation should be controlled in conjunction with other specialties, and the operation of ventilation equipment should be controlled according to unified disaster prevention requirements. Design information shall be provided according to the requirements of the monitoring profession.

(6) Cooperation with operation management. Ventilation equipment cannot be installed once and for all, need daily maintenance and repair, need to put forward daily maintenance

management requirements to the operation management department, and staff requirements.

 Exercise

7.1 There are several aspects in the study of railway tunnel fire both at home and abroad.

7.2 What are the rescue and evacuation facilities for railway tunnels and what are the requirements to be met?

7.3 What are the methods of ventilation equipment arrangement for railway tunnels? Briefly describe the applicable conditions of each method.

 Key vocabulary:

critical wind speed 临界风速
evacuation outlet 疏散出口
prevention scheme 预防方案
wind speed distribution 风速分布

emergency rescue station 紧急救援站
hoisting arrangement 吊装布置
turbulence coefficient 紊流系数

Chapter 8 Metro Disaster Prevention and Rescue

[Important and Difficult Contents of this Chapter]
(1) Causes and characteristics of metro fire.
(2) Automatic fire alarm system of metro.
(3) Emergency evacuation passageway technology of metro.

8.1 Overview of Metro Fires

8.1.1 Metro Fire Accident Cases

Metro is the main artery of urban traffic and one of the most important means of transportation for urban residents. Because the metro environment is closed. Dense, ventilation and fume exhaust facilities and evacuation escape space is limited. Once equipment and facilities failure, fire, explosion and other accidents happened, emergency treatment will be very difficult. With the development of urban metro system, sudden disasters such as fires and explosions pose a great threat to the operation of the metro system. They may not only cause casualties and property losses, but also cause great damage to urban functions and form a serious social impact. Table 8-1 lists some cases of urban metro system fires both at home and abroad.

Some domestic and foreign urban metro fire cases   Table 8-1

NO.	Metro station name	Fire (as a disaster) time	Causes	Damage
1	Shepards Bush-Holland Park, UN	1958	Malfunction of electrical equipment	1 person died, 51 were injured
2	Tokyo Metro Ribi Valley Line Raupongu Station Shinjiachi Station,	1968	The equipment of the running train caught fire	11 person injured, 3 compartments burned
3	Beijing Metro Line1, China	1969	Fire due to electrical equipment failure	6 people died, more than 200 people were injured; 2 electric locomotives were burned down
4	Montreal Metro, Canada	1971	Short circuit fire caused by train collision with tunnel	1 person died; 24 carriages were burnt down; a total loss of $5 million

continued

NO.	Metro station name	Fire (as a disaster) time	Causes	Damage
5	Paris Metro Line 7, France	1973	Artificial arson in the carriage	2 people died of suffocation; the carriage was burnt down.
6	Bostonmetro, USA	1975	Short circuit of tunnel power supply	34 people were injured.
7	Toronto Metro, Canada	1976	Artificial arson	4 carriages burnt down
8	Lisbon Metro, Portugal	1976	Fire due to technical malfunction	4 carriages burnt down
9	Cologne Metro, German	1978	Unextinguished cigarette butts dropped on the rear bogie chassis causing a fire	Tram and electric track burned down; 8 people were injured
10	Philadelphiametro, USA	1979	Short circuit of tunnel power supply	148 people were injured; a carriage was burnt down
11	New Yorkmetro, USA	1979	Discarded unextinguished cigarette butt ignited the tank.	4 people were injured; 2 carriages were burnt down
12	San Franciscometro	1979	Fire caused by damage of transverse current collector	1 person died and 54 were injured; 5 carriages were burnt down and 12 carriages were damaged.
13	Hamburg Metro, Germany	1980	Arson on car seats	4 people were injured; 2 carriages were burnt down
14	Moscow Metro, Russia	1981	Short circuit of tunnel power supply	7 people died; 15 carriages were burnt down.
15	Bonnmetro, Germany	1981	Technical failure	Tram burnout.
16	New York Metro (Okot Brickayer Station)	1981	Electronic malfunction caused the train to fire	7 people died; 2 carriages were burnt down.
17	New Yorkmetro	1982	Fire caused by control gear failure	86 people were injured; 1 carriages were burnt down
18	Munich metro, Germany	1981	Electrical failure	56 people were injured; 1 carriages were burnt down
19	Hamburg Metro, Germany	1982	Arson on car seats	4 people were injured; 2 carriages were burnt down
20	New Yorkmetro	1985	Artificial arson	15 people were injured; 16 carriages were burnt down

continued

NO.	Metro station name	Fire (as a disaster) time	Causes	Damage
21	Tokyometro, Japan	1985	Fire caused by damage and fever of lower bearing of train	Some carriages were burnt down
22	Berlin Metro, Germany	1986	Technical failure	Carriages were burnt down
23	King's Cross on London Metro, UK	1987	Passengers litter cigarette butts in the gap between wooden escalators, causing a fire in a machine room under the escalator.	32 people died; more than 100 people were injured
24	Zurich Metro, Switzerland	1991	Short circuit of cables in the car	Some carriages were burnt down
25	Bakumetro, Azerbaijan	1995	Engine electrical aging and short circuit	558 people died; 269 people were injured
26	Daegu metro line 1, South Korea	2000	Artificial arson	192 people died; 147 people were injured
27	Beijing metro Chong wen men station, China	2000	fume and fire broke out in the exhaust fan of the running train carriage	The station was closed for 50 minutes, and the top of the tunnel was fumed and blackened without casualties
28	Simpland Station on Line 13, French	2005	One metro train caught fire and then spread to the opposite metro train	12 people were injured; Line 4 was partially closed
29	Chicagometro, USA	2006	The train derailed in the tunnel and the carriage caught fire.	152 people were injured
30	Kiev metro Ozakoga station, Ukraine	2012	The ceiling was set ablaze by a pendant lamp and the fire spread rapidly	
31	Moscow Metro Line 1	2013	A fire broke out between the "Huntsman mall" and the "Lenin Library" station	About 4,500 passengers were evacuated; 15 were injured

8.1.2 Causes and Characteristics of Metro Fire

Through the analysis of metro fire accidents at home and abroad, Figure 8-1 gives the cause statistics of metro fire. It can be seen that the main causes of metro fires are:

(1) The failure of the mechanical and electrical equipment of the vehicle itself causes the fire of the vehicle. The metro fire accidents caused by vehicle fire caused by vehicle mechanical and electrical equipment failure account for 40.43% of the total investigation cases, which is the main cause of fire. The aging of several parts of the vehicle and the short circuit are the causes of the fire.

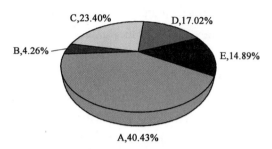

A—Vehicle accident due to mechanical and electrical equipment failure of the vehicle itself;
B—Traffic accidents such as vehicle impacts;
C—Arson, etc;
D—Ageing and Failure Accidents of Cables and Electrical Equipment in Tunnels;
E—Other reasons.

Figure 8-1　Statistical table of metro fire causes

(2) Arson and other human factors: Arson, cigarette butts and other human factors are one of the main causes of metro tunnel fire. Human-induced metro tunnel fires accounted for 23.4% of the total investigation cases. Typical cases were the fire at King's Cross Station of London Underground Tunnel on November 18, 1987 and the fire at Grand Central Road Station of Daegu Underground Line 1 on February 18, 2003.

(3) Aging and failure of mechanical and electrical equipment in section tunnel. There is a certain amount of electrical and line equipment in the interval tunnel. Because the interval tunnel is generally dark and humid, electrical and line equipment will inevitably age and malfunction during use, resulting in interval tunnel fire. In the actual metro fire cases, 17.2% of the fires are caused by such reasons.

The metro tunnel is crowded with people, the space is relatively narrow, and the evacuation and rescue is difficult. The fire has the following characteristics:

(1) It is difficult to exhaust fume and dissipate heat, and the temperature is high. As there are few connecting channel with that outside, once a fire occurs, the heat generate by the combustion is difficult to dissipate, and the temperature rises quickly and flashover is easy to occur quickly. The results show that the flashover time is 5-7 minutes after the fire.

(2) High temperature flue gas is seriously harmful. Because of less entrance, poor air circulation, inadequate ventilation, inadequate oxygen supply, incomplete combustion will occur in the fire, resulting in the rapid rise of toxic gases such as carbon monoxide concentration, leading to serious casualties. In addition, with the diffusion and flow of high-temperature fume, it will also lead to a serious reduction in visibility, affecting the evacuation of personnel and firemen

to put out the fire.

(3) Difficulty in evacuation. Due to limited conditions, the metro has few entrances and exits, and the evacuation distance is long. In case of fire, the flow direction of high-temperature fume at the entrance and exit is the same as the escape direction of personnel, and the diffusion and flow speed of fume is much faster than the evacuation and escape speed of the crowd. People will escape under the cover of high-temperature fume, resulting in reduced visibility and panic in the crowd. At the same time, some gases in the fume, such as NH_3, HF and SO_2, will make people's eyes can not open, people will be more frightened, may collapse to the ground or blind escape, causing unnecessary casualties.

(4) Difficulty to put out the fire. When a metro fire occurs, because of the closed environment, the fire location and real-time information of the fire scene are difficult to be accurately known, and the fire extinguishing route is few, which makes the fire extinguishing extremely difficult.

8.2 Automatic Fire Alarm For Metro

8.2.1 Fire Alarm System

Fire Alarm System (FAS) is designed according to the national standards *Code for Design of Metro* (GB 50157—2013) and *Code for Design of Automatic Fire Alarm System* (GB 50116—2013), and according to the actual situation of metro operation and management. Generally, the design method of two-level management and three-level control is adopted, that is, two-level management method of central level and station level (station, depot, parking lot) is adopted, and the FAS of the whole line is an independent monitoring and management system, which is not integrated with other systems. The Operation Cooperation Center (OCC), backup center and maintenance center are at the central level, and the fire control rooms at stations, depots, parking lots and training centers are at the station level. The central level is the dispatching center of the whole line fire alarm system, which has the right to monitor, control and manage the information of the whole line fire alarm system and the fire control facilities. The station-level jurisdiction covers the station and its adjacent half sections, depots, parking lots and other areas. The station-level can realize automatic monitoring and control of FAS system equipment in the station or jurisdiction, and realize automatic management of fume control and exhaust, firefighting, evacuation and disaster relief equipment. The control of disaster prevention equipment (ventilation, water supply and drainage, lighting, escalator, fire curtain, gas extinguishing and other equipment) in FAS system of the whole line can realize the three-level control method of central control level of disaster prevention command center, station level of station disaster prevention control room and equipment site control level.

Chapter 8 Metro Disaster Prevention and Rescue

The latest national standard *Code for Design of Metro* (*GB 50157—2013*) defines the automatic fire alarm system as an automatic management system for fire prevention and relief work of metro, which includes metro fire alarm, fire control and other monitoring of metro fire disaster and linkage control of firefighting equipment. Article 19.2.1 stipulates that the automatic fire alarm system shall have automatic fire alarm, manual fire alarm, connecting and network information alarm, etc., and shall realize the control of fire rescue equipment and linkage control with relevant systems. Article 19.2.2 the automatic fire alarm system shall be composed of the central monitoring and management system installed in the control center, the station-level monitoring and management system of the station and the vehicle base, the field-level monitoring and control equipment and the relevant connecting network.

The central monitoring and management system of automatic fire alarm system is generally composed of operator workstation, printer, connecting network, uninterruptible power supply and display screen. The station level of automatic fire alarm is generally composed of fire alarm controller, graphic display device of fire control room, printer, uninterruptible power supply and fire control linkage controller, manual alarm button, fire telephone and field network. The information transmission network of automatic fire alarm and linkage control in metro line should use the metro public connecting network, and the site-level network of automatic fire alarm system should be configured independently.

8.2.2 Design of Fire Alarm System for Typical Metro Engineering

1. Engineering overview

The fire alarm system of Shanghai Rail Transit Line X is designed according to the control center level and station level two-level monitoring and management method. The first level is the control center level, which carries out centralized monitoring and management of the disaster alarm system of the whole line of Shanghai Rail Transit Line X. The second level is the station level, which monitors and manages the fire control equipment of the disaster prevention and alarm system within the jurisdiction of the station level (vehicle base). The main substation shall be equipped with a regional fire alarm controller, which shall be included in the management of adjacent stations. The system is interconnected with Integrated Supervisory Control System (ISCS) at control center level and station level respectively.

2. Set up specifications and standards

The design specifications and standards adopted in this system mainly include *Technical Code of Urban Rail Transit* (*GB 50490—2009*), *Code for Design of Metro* (*GB 50157—2013*), *Code for Design of Zhejiang Province Urban Rail Transit* (*DJG 08109—2004*), *Technical Standards for Shanghai Urban Rail Transit Engineering* (*Implemented*) (*STB/ZH-000001—2012*), *Code for Installation and Acceptance of Fire Alarm System* (*GB 50166—2007*), *Standard for Design of*

Intelligent Building (GB 50314—2006), *Technical Code for Remote-monitoring System of Urban Fire Protection* (GB 50440—2007), *Code for Design of Electronic Information System Room* (GB 50174—2008), *Code for electrical design of civil buildings* (JGJ 16—2008), *Fire-fighting Linkage Control System* (GB 16806—2006) and other relevant national and Shanghai design codes, codes and standards. If two standards do not meet, the higher standard shall apply.

3. Design principles

The fire alarm system is designed according to the principle of two-level management and three-level control. The whole system is composed of central monitoring management level, station (each station, intermediate air shaft, main substation, depot/parking lot, control center building) monitoring management level, field control level, relevant network and connecting cooperation, etc., which are set in the control center. The fire alarm system is based on safety, reliability and practicality, and embodies the guiding ideology of "people-oriented" design. The fire alarm system shall implement the principle of "prevention first, combination of prevention and elimination", comply with the relevant national regulations and norms, and conform to the relevant provisions of Shanghai Fire Bureau. FAS system is designed according to a fire which occurs in the same time of the whole line.

Underground stations, underground sections, control center buildings, main substations, large garages of depots/parking lots, inspection and repair garages, important material banks and other important buildings shall be designed according to the first-class protection object of fire alarm. The general production and office buildings of ground and elevated station, depot/parking lot shall be designed according to the secondary protection object of fire alarm.

The control center level and station level two-level management method are adopted in the line change fire alarm system. The control center level realizes the centralized monitoring and management of the automatic fire alarm system of the whole line. The station level sets fire alarm controllers in each station, depot/parking lot and main substation, which can independently carry out the monitoring and management of fire control within the scope of its management. The fire control center of the whole line is located in the control and command center, and the disaster prevention and command centers of the station, depot/parking lot and other levels are located in the control room of the station and the control room of the depot/parking lot complex building respectively. The normal operation method and the accident operation method are adopted for the fan and the air valve which are shared by the fume prevention and exhaust system and the supply and exhaust system. The normal operation method is monitored and managed by the equipment monitoring system in real time, and the accident operation method is controlled by the fire alarm system to the building automatic system (BAS). After receiving the instruction, BAS starts the relevant fire method according to the content of the instruction, realizes the fire method control of the relevant equipment, and feeds back the execution signal of the instruction, which is displayed on the disaster relief command screen to help the development of the disaster relief command. The control equipment of fire water pump and special fume exhaust fan shall be controlled by bus coding module, and emergency manual direct control device shall also be installed in the fire control room. The emergency manual direct control device is uniformly set up by the integrated backup panel (IBP) of the equipment monitoring system. The fire-fighting broadcast system and station broadcast system are used together, and the fire emergency broadcast function is provided. In case of fire, the emergency broadcast can be forcibly transferred to the emergency broadcast state. The connecting system such as depot/parking lot does not have a public address, and fire alarm or alarm bell shall be set by the system.

Design labeling and main parameters include:

Central control response time of control center: < 2 s; Central information response time of control center: < 2 s; Site control response time: < 1 s; Site information response time: < 1 s; Response time of fire alarm loop: < 0.85 s; Mean barrier-free time (MTTR) of main equipment of fire alarm system: < 30 min; The length of each bus loop shall not be less than 1 500 m under the condition that the cross section of the loop conductors is 1.5 mm2; Ground resistance: $\leqslant 1\Omega$; Overall service life of the system: 15 years.

4. Network composition

FAS full-line network adopt peer-to-peer ring network structure, The fire alarm systems at the control center level, each station, depot/parking lot, main substation, intermediate air shaft, control center building, etc., all serve as nodes on the FAS network, Each fire alarm controller has the same status in the network connecting, each node can independently complete the monitoring and control of the equipment in the jurisdiction area, each node is equal to each other, if there is short circuit, open circuit or failure between nodes, nodes will be automatically isolated, network connecting will not be interrupted. Dedicated optical fiber is used as the connecting channel of the system. The alarm host of the control center connects each node in a ring through 6 optical fibers (4 for 2 standby) in the optical cable provided by the connecting system to realize network connecting.

The main substation (intermediate air shaft) is a certain distance from the adjacent station, and the relative monitoring points are less. There are two schemes for the main substation to access the system: The invention relates to a multiple-view screen scheme, which takes a main substation (intermediate air shaft) as a loop of an adjacent station and connects the fire alarm controller of the adjacent station through a cable or an optical cable, and a remote multiple-view screen is arranged in the main control room of the main substation (intermediate air shaft) to display the alarm and linkage information in the main substation (intermediate air shaft). The other is an alarm scheme, in which a fire alarm controller is arranged in the main control room of the main substation (intermediate air shaft), and the main substation (intermediate air shaft) is connected as a network and connected to the connecting network of the FAS system.

5. Control center level composition

The control center communicates with each station level (including station, main substation, intermediate air shaft, depot/parking lot and control center building) FAS, which can monitor the operation status of the whole line of firefighting equipment, receive and display the alarm signals sent from each station level, record and print automatically, and manage the historical archives; Issue disaster prevention and relief instructions to the disaster prevention and control rooms of each station, organize, coordinate, direct and manage the whole line of disaster relief work, timely report the disaster situation to the relevant superior fire department, and regularly output various data and reports; Receive the information of the master clock to synchronize the fire alarm system with the master clock.

The central fire alarm controller is set up in the dispatching hall of the control center as the main control unit of the whole line fire alarm. The fire alarm system at the central level is equipped with independent graphic display terminal, and the Cooperation with the comprehensive display screen (large screen) of the dispatching hall is realized through FAS workstation, and the fire alarm information is sent to the comprehensive display screen.

6. Station-level system composition

The station control level monitors and controls all kinds of disaster prevention equipment in the station level (including main substation, intermediate air shaft, depot/parking lot, control center building) and the area under its jurisdiction, receives fire alarm signals between the station and the area under its jurisdiction, and displays fire alarm and fault alarm positions. Report the disaster to the control center, accept the instruction from the control center, start the relevant firefighting equipment to put into operation in fire method, and organize and lead the evacuation by means of connecting tools.

The monitoring and management of the station is arranged in the station control room of each station, The monitoring and management of depot/parking lot is set in the duty room of comprehensive building, the monitoring and management of main substation is set in the main control room of main substation, the monitoring and management facilities of intermediate air shaft are set in the duty room or control room, and the monitoring and management of control center building is set in the fire control duty room of control center building.

The station monitoring and management level shall independently perform the monitoring and management functions of the FAS system within its jurisdiction. Station monitoring and management level consists of fire alarm controller, fire trigger (including fire detector and manual alarm button, very early fume detection alarm, etc.), fire alarm device, fire control linkage controller, quasi-terminal display equipment, fire telephone host, printer and other equipment.

The fire alarm controller is connected with the equipment monitoring equipment through the bi-directional connecting Cooperation to complete the linkage control of the dual-purpose annulus equipment. At the same time, the fire alarm controller sends the information to the control center through the optical fiber media provided by the connecting system.

Set a direct start button on the integrated backup panel (IBP) set up on the station console for the operation of important firefighting equipment. Important firefighting equipment includes:

hydrant pump, spray pump, high-pressure water mist pump, fume exhaust special machine and so on. The direct start button of the integrated backup disc can start the important firefighting equipment directly without passing any intermediate equipment in case of fire. In terms of level, this is the most advanced linkage equipment. At the same time, the operation and failure status of the important firefighting equipment, as well as the position and status of the start button can also be displayed on the comprehensive backup plate.

Take each depot (main substation, depot/parking lot, large control center) as a unit, set up independent fire-fighting private telephone network. Set up special direct telephone switchboard for fire control in station control room (signal building duty room for depot/parking lot, main substation main control room and fire control duty room for control center building). Fixed fire extension telephones shall be installed in important places such as substation rooms, fire pump rooms, environmental control electronic control rooms, gas fire extinguishing operation panels and elevator rooms. Firefighting telephone jacks shall be arranged at appropriate positions (such as manual alarm buttons, beside fire hydrant buttons) in the platform layer, hall layer and underground section of the station, so as to realize voice connecting between the station control room and these places.

7. Depot/parking system composition

The fire alarm controller is set up in the fire control room of the comprehensive building of the depot/parking lot as the station-level fire alarm controller, and is directly connected with the whole line fire alarm system. A regional fire alarm controller shall be set up in the parking train inspection depot and the overhaul combined depot, and the fire control management functions thereof shall be managed in the control room of the comprehensive building. Various detectors shall be installed in equipment rooms and management rooms such as hybrid substation, comprehensive building, overhaul warehouse, general material warehouse, operation warehouse, combined garage, etc.

The fire control room of the comprehensive building in the parking lot shall be equipped with the main fire control telephone, and the extension and jack of the fire control telephone shall be set up in the comprehensive building, the maintenance depot, the hybrid substation and other places.

The fire alarm controller of the complex building is connected to the whole line fire alarm system through optical fiber and fire alarm controller of each station and control center.

Firefighting broadcast shall be set up in the operation depot and maintenance depot of depot and parking lot, and the broadcast console shall be set up in the firefighting duty room. The fire emergency broadcast shall conform to the fire alarm system specification.

The FAS sets alarm bells in other areas of the depot/parking lot.

The fire alarm system of the parking lot is composed of the fire alarm controller, graphic monitoring terminal of the comprehensive building, the regional fire alarm controller of the parking lot inspection depot, the overhaul joint depot, various detectors within the jurisdiction, manual alarm buttons, input and output modules and other field equipment.

8. Main substation system composition

The main substation is equipped with a vehicle-level fire alarm controller, which is directly connected with the whole line fire alarm system through optical cables at the adjacent station. The fire alarm range of the main substation is connected to the cable passage way adjacent to the station. Fixed fire-fighting telephones shall be set up in the main substation, and the fire-fighting telephones shall be connected to the adjacent stations through telephone lines. A main substation fire alarm system is composed of a main substation fire alarm controller, a graphic monitor terminal, a lattice detector within that jurisdiction, a manual alarm button, an input/output module and other field devices,

9. Middle air shaft

Station-level fire alarm controllers are set up in the air shaft of the section, which are directly connected with the fire alarm system of the whole line through optical cables in the adjacent stations, and their fire control monitoring and management functions are respectively hosted in the nearby stations. The graphical monitoring terminal in the vicinity of the station shall be able to display the fire alarm system information of the section substation and the section air shaft. The zone substation and zone air shaft fire alarm system are composed of zone air shaft fire alarm controller, various detectors, manual alarm buttons, I/O modules and other field equipment within the jurisdiction. Power supply equipment shall be set up in the section air shaft, including double power self-cutting box, uninterruptible power system UPS and distribution box, which shall be used together with Electronic and Mechanical Control System (EMCS) and Access Control System (ACS) (one distribution loop shall be provided to EMCS and ACS respectively).

10. Control Centre Building

The dispatching hall of the control center shall be equipped with a central-level fire alarm controller (host computer), and the matching color graphic monitoring terminals shall be arranged on the dispatching platform for environmental control and disaster prevention. The dispatching station shall be made and arranged in a unified manner by the control center, and the broadcasting control box, closed circuit television display terminal, service telephone, local direct line telephone, fire-fighting radio telephone and environment-controlled dispatching switchboard shall be set up in the dispatching station by the connecting system. The equipment monitoring operation terminal is set up by the equipment monitoring specialty.

The central level alarm system is connected with the central level electromechanical equipment monitoring system through the connecting Cooperation to realize the transmission of information. In case of fire at each station, the comparison table between the actual operation and the actual operation of all the monitoring equipment of FAS and EMCS (i.e., the complete fire linkage operation object) shall be displayed on the graphical monitoring terminal Cooperation. FAS is equipped with a separate laser printer to print real-time information and various reports. The central level equipment of FAS shall be supplied by centralized UPS, and the terminal distribution box shall be set up by this specialty. The terminal distribution box provides both EMCS and ACS with operation power.

11. Transfer station

According to the operation and management requirements of "one station master, one set of teams, resources sharing and area control" for shared station, only one set of fire alarm system shall be set up in principle for shared station according to the design principle of "building before building" during construction, which shall be implemented by the first-built line project, and the control scope shall include all areas of the two-line station. According to the principle of one-time design and step-by-step implementation, the equipment of the back-to-back line shall be purchased and connected to the system implemented in the current phase during the later stage of the project implementation.

12. Main functions of the system

1) Main functions at the control center level

Monitor and control the fire alarm equipment and special firefighting equipment in the whole line. Monitor, display and record the operation status of all firefighting equipment in the whole line; When the controlled equipment fails or the status changes, sound prompts are sent and printed to record the time and place of occurrence.

Receive fire alarm, fault alarm and work status information of disaster prevention equipment from FAS system at station level (each station, main substation, intermediate air shaft, depot/parking lot, control center building). When a fire alarm occurs, the alarm point shall be displayed on the color graphic monitoring terminal with a map-type picture in a timely manner, the alarm time and place shall be printed, the acousto-optic alarm signal of the fire alarm shall be activated, and the fire confirmation time of the dispatcher shall be displayed.

Organize and direct the fire rescue work of the whole county, select the predetermined solution, issue fire rescue orders and safety evacuation orders to the station level, and direct the development of disaster relief work. In case of fire in the underground tunnel, coordinate the control conditions of two adjacent tunnels and issue control orders to the station. Receive the information of the master clock to synchronize the fire alarm information with the time system. Establish a database and file management, regularly output all kinds of data and reports.

The fire alarm system of the whole line shall be subject to operation authority management. There are multi-level passwords, and operators of different levels should have different access and operation permissions. The control center level has the highest operational authority. It can edit the operators of each site on-line and download the program, and modify the field parameters. After the parameter setting is modified, it is downloaded to the alarm controller of each station through the network.

The control center level can confirm the disaster situation of the monitoring site by operating the keyboard and display terminal of the television monitoring system (CCTV). According to the actual situation of the fire, fire control and rescue orders and safety evacuation orders shall be issued to the relevant areas, and the whole line of disaster prevention and rescue work shall be commanded by means of connecting tools such as whole line disaster prevention and dispatching telephone, outside line telephone, closed circuit telephone, train radiotelephone, etc. The fire conditions shall have priority.

2) Main functions at station level

Station level includes station, intermediate air shaft, depot/parking lot, main substation and control center building, etc. Receive the fire alarm signal between the station and the area under its jurisdiction, and display the fire alarm or fault alarm position. Monitor the operation status of various fire alarm equipment and special firefighting equipment in the station and the area under its jurisdiction. Confirm the disaster situation and report to the control center and relevant departments for connecting and transmission of fire information. Receive fire rescue orders and evacuation orders issued by the fire control center, organize and induce passengers to evacuate safely. After confirming the fire, instruct the station-level equipment monitoring system to operate according to the predetermined fire method. Through the cooperation between FAS system and other systems, the related equipment will operate according to the fire condition.

8.3 Emergency Evacuation Passage Technology for Metro

Metro is called "green traffic" because of its safety, comfort, large capacity, fast, punctuality, low energy consumption and less pollution. It is more and more popular and greatly improves the problem of urban traffic congestion. Because the construction, equipment and operation of the metro are underground, with a large number of mechanical and electrical equipment and a certain amount of flammable and combustible substances, there are many passengers and staff in the process of operation, so there are many potential fire factors. Metro fire has the following characteristics:

(1) Flue gas diffuses rapidly. The internal space of the MTR is relatively closed, so the tunnel is narrower, so the fume is difficult to diffuse and will fill the station and tunnel sections.

(2) Poor escape conditions. Mainly in the vertical height of the deep, escape route is few, escape distance is long.

(3) The time allowed for escape is short. Tests show that the time allowed for passengers to escape is only about 5 minutes. In addition, once the clothes of the passengers ignite, the fire will expand in a short period of time, the time allowed to escape is even shorter. The allowable escape time in *Code for Design of Metro* (*GB 50157—2013*) is 6 minutes.

(4) Difficulty to prevent arson. Metro personnel mobility, coupled with a lot of ventilation openings, so the metro arson incident suddenly strong, in the absence of precursors, passengers are difficult to arouse vigilance, take precautionary measures in advance.

Chapter 8 Metro Disaster Prevention and Rescue

(5) Difficulty in evacuation and shelter. Metro stations and tunnels have narrow spaces and few entrances and exits, but they are very crowded during rush hours. In case of fire, passengers are apt to panic, crowd each other and collapse, tread injury or tread death in the absence of command. In addition, when the metro fire, the direction of escape and the direction of fume diffusion are from the bottom to the top, the entrance and exit of personnel may be the entrance and exit of fume, increasing the difficulty of evacuation.

Aiming at the problems of evacuation and rescue in the above-mentioned metro station fire, This section presents a method for setting up emergency escape passages for metros. By setting up emergency evacuation escape passages, which allows passengers to escape quickly and safely in emergency situations, providing an additional escape route for passengers in metro cars to flee quickly. At the same time, the emergency evacuation escape passage can also be used as a temporary shelter for the aged and frail who have difficulty in moving and the injured who are likely to appear in the accident, and can be used as a fast passage for the firefighters to enter the scene for fighting and rescuing. After the passengers are evacuated effectively, the emergency evacuation escape passage can also be used as part of the ventilation and fume exhaust system, which can improve the flexibility of fume control in case of fire. And in peacetime emergency evacuation escape channel can also be used as a rapid way to deal with emergencies, transport materials.

As shown in Figure 8-2 to Figure 8-4, the emergency evacuation escape passage is divided into two layers: the station hall layer and the platform layer. The platform layer is provided with a plurality of passage entrances on the inner side wall of the station hall layer, and a fire-proof rolling shutter door is arranged at the entrance of the passage. The platform layer and the station hall layer are provided with evacuation stairs, the evacuation stairs in the platform layer lead to the station hall layer, the corresponding entrance is arranged on the station hall layer, and the stairs in the station hall layer lead to the ground.

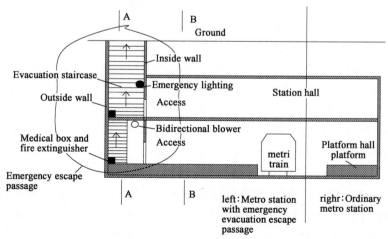

Figure 8-2 Schematic diagram of emergency evacuation escape passage

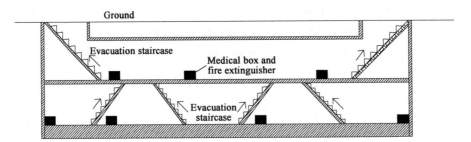

Figure 8-3 Schematic diagram of emergency evacuation escape passage (section A-A)

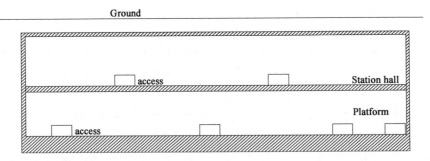

Figure 8-4 Schematic diagram of emergency evacuation escape passage (section B-B)

Inside the platform layer and the station hall layer, there are also two-way connecting machines, emergency lighting, medical boxes and fire extinguishers.

The fire curtain door at the entrance of the emergency evacuation escape passage is normally closed. Once the fire occurs and the personnel need to evacuate, the fire curtain door is opened, and the personnel in the station can enter the emergency evacuation escape passage for shelter and treatment, and escape to the ground directly through the stairs. Firefighters can also enter the station directly from the ground through the evacuation staircase for firefighting and rescue work.

The two-way ventilator is turned on when the fire occurs in the station, so that the air pressure in the emergency evacuation escape passage is higher than the external air flow, thereby effectively preventing the high-temperature fume from spreading into the emergency evacuation escape passage and threatening the life safety of the personnel in the emergency evacuation escape passage. The fire extinguisher can be used to put out the fire quickly when the fire breaks out in the passageway to ensure the safety of the people in the passageway. Emergency lights provide temporary lighting. Medical equipment and medicines in medical kits can be used for emergency treatment of elderly and frail persons with limited mobility and those injured in the accident after they enter the emergency evacuation passageway.

As shown in Figure 8-5, for metro stations with side platforms, emergency evacuation escape passages are provided on the outside of both platforms. The outer side wall of the emergency evacuation escape passage can be shared with the side wall of the metro station, and the inner side wall of the emergency evacuation escape passage is a firewall, which separates the emergency evacuation escape passage from the platform of the station.

Chapter 8 Metro Disaster Prevention and Rescue

Plane position of emergency evacuation escape passage

platform(side)
metro train
metro train
platform(side)

Plane position of emergency evacuation escape passage

Figure 8-5 Schematic diagram of plane position of emergency evacuation escape passage of side platform

As shown in Figure 8-6, for a metro station using an island platform, the emergency evacuation escape passage is located on both sides of the platform so that the emergency evacuation escape passage is separated from the platform by the tracks of the train. In case of emergency, the door on the track can be opened, and the personnel on the side platform can pass the train into the emergency evacuation escape passages on both sides of the station.

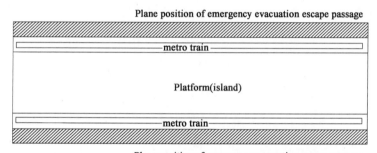

Figure 8-6 Schematic diagram of plane position of emergency evacuation escape passage of island platform

The space of emergency evacuation escape passage can be reserved in the excavation process of metro station foundation pit, that is, to enlarge the dimensions of both sides of the foundation pit, and the additional part is reserved by the passage. The retaining structure (diaphragm wall) of the station foundation pit can be used permanently as the outer side wall of the passageway.

 **Exercise**

8.1 What are the main causes of urban metro fires?
8.2 What are the characteristics of urban metro fire?
8.3 What is the design pattern of two-level management and three-level control?
8.4 Briefly describe the function of emergency evacuation escape passage of metro.

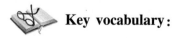 **Key vocabulary:**

building automatic system   建筑自动控制系统

central monitoring and management system 中心监测管理系统
fire alarm system 火灾预警系统
integrated supervisory control system 综合监控系统
integrated backup panel 综合后备盘
mean barrier – free time 平均无障碍时间
uninterruptible power system 不间断电源系统

Chapter 9 Disaster Prevention and Rescue of Other Underground Space

[Important and Difficult Contents of this Chapter]

(1) Fire risk, fire prevention measures and firefighting and rescue measures for underground shopping malls.

(2) Fire risk of underground parking lot, fire prevention measures and firefighting and rescue measures.

(3) Mine fire hazards, fire prevention measures and firefighting and rescue measures.

(4) Fire characteristics, fire prevention measures and firefighting and rescue measures of civil air defense project.

9.1 Disaster Prevention and Rescue of Underground Shopping Mall

The expansion of the internal scale and scope of use of underground shopping malls speeds up the development of urban underground space, but at the same time it brings convenience to people, its potential fire risk has also increased significantly. Therefore, an objective understanding of the fire characteristics of underground shopping malls and a targeted study of its disaster prevention and rescue work are of great practical significance to do a good job in fire control, maintain social development and ensure the safety of people's lives and property.

9.1.1 Fire Hazard of Underground Shopping Mall

1. High turnover of personnel

Underground shopping malls are places where large numbers of people move. Especially in recent years the rise of underground supermarkets such as the nature of the shopping malls, the daily in and out of the very large number of people. Especially in some big cities in the large shopping malls, every day in and out of all kinds of people are more than a hundred and thousand, or even more. In addition, in some special days, such as holidays, businesses to carry out a variety of promotions, that is the crowds of customers, countless. Such a crowded place, once the fire happens, will bring great difficulties to the rescue and evacuation work.

2. More combustible goods

Underground shopping malls deal in mostly combustible goods. Most of these goods are displayed in bulk or stacked on shelves and counters while some commodities, such as clothing and hats, various textiles, arts and crafts, luggage and so on, are still hanging in the air. Even if some commodities are made of non-combustible materials, their packing cases and boxes are all combustible. According to these characteristics, the combustible load of underground shopping malls is much greater than that of any other places. Once a fire breaks out, it can burn rapidly and violently.

3. Difficulty in lighting and ventilation

Difficulties in ventilation and lighting are inherent characteristics of underground engineering. But it should have such essential conditions as the material circulation, the personnel movement place, the good air circulation and the sufficient light. To overcome the innate deficiencies, massive ventilation and the day lighting facilities are needed. But the facilities require huge amount of electricity. It is recognized as a disaster-causing factor of consuming large amount of electricity in places with large numbers of combustible materials.

4. More electrical equipment

Because underground shopping malls must rely on artificial lighting, the underground shopping malls are equipped with a large number of fluorescent lighting fixtures. In addition, a number of temporary power sockets are installed in the places where the shopping malls operate lighting equipment and household appliances for testing purposes. In the holidays, underground shopping malls inside and outside the temporary installation of a variety of lanterns, adding festive atmosphere. Therefore, underground shopping malls have the characteristics of many electrical equipment, a wide variety of complex circuits and long service life. If the design, installation and use of careless, it is very easy to cause fire accidents.

9.1.2 Fire Prevention and Control of Underground Shopping Mall

1. Fire compartmentation

In order to effectively contain the spread of fire, reduce the loss of fire, and carry out firefighting and rescue, underground shopping malls should be strictly in accordance with the relevant norms of fire compartmentation. According to the requirements of *Code for Fire Protection Design of Civil Air Defense Works* (*GB 50098—2009*), the maximum allowable floor area of each fire prevention zone shall not be greater than 500 m^2. When the automatic fire extinguishing system is installed, the maximum allowable building area can be increased to 1000 m^2. When the automatic fire alarm system and automatic fire extinguishing system are installed, and the decoration materials of Grade A are used, the maximum allowable building area of the fire prevention zone should not be greater than 2000 m^2. However, in practice, the decoration materials of underground shopping malls are not up to Grade A decoration materials. Therefore, the maximum area of each fire prevention zone is 1000 m^2. When the area of the local shopping mall is larger than 20000 m^2, the adjacent areas shall be divided by fire compartments, the walls of which shall be firewalls, and the doors on the firewalls shall be normally open Class A fire doors that can be closed by themselves.

2. Fume prevention and exhaust

When the local shopping mall fire, a large number of toxic fumes in the limited space wantonly spread, very easy to cause a large number of deaths, so the fume control design of underground shopping malls is more important. You must first ensure that each fire zone is equipped with a fume outlet. The ideal way to prevent and exhaust fume is to prevent fume from entering the evacuation passage or the shelter strictly and ensure the absolute safety of these areas. At the same time, it can allow free access to these areas for victims and firefighters. The current fume control and exhaust methods cannot meet the two requirements at the same time, but the air curtain fume control and exhaust methods can meet the two requirements very well. In order to achieve better ventilation and fume exhaust effect, the blowing and sucking air curtain should be adopted to prevent fume exhaust in underground shopping malls.

3. Fixed fire extinguishing and automatic alarm system

Because of the particularity of the underground shopping mall and the danger of the fire in this area, the fire extinguishing effect of the local shopping mall will be very limited if we want to rely on the firemen to extinguish the fire. Therefore, in this place, the fixed fire-fighting facilities and automatic alarm system must be designed to achieve the purpose of self-help. When there are potential fire hazards, the design can be timely found, alarm in advance, and effective remediation, so as to minimize the harm. At present, fire hydrants and automatic sprinkler systems are commonly used as fixed fire extinguishers in underground shopping malls.

4. Non-combustible and difficult-flammable interior decoration

In order to reduce fire and control the spread of fire, the interior decoration of underground shopping malls should be non-combustible and difficult-flammable as far as possible. It is required that the interior decoration should use less difficult-flammable materials, and the use of flammable materials is strictly prohibited. Therefore, in the interior decoration design of underground buildings, the requirements of *Code for Fire Prevention in Design of Interior Decoration of Buildings* (*GB 50222—2015*) shall be strictly observed. Class A decoration materials shall be adopted for the business hall, evacuation aisle and foyer ceiling, floor, wall surface, sales counter, fixed shelves and exhibition stand of the shopping mall, and Class B1 decoration materials shall be adopted for the partition. In addition, the decoration is not allowed to block the fire protection facilities.

5. Safe evacuation

The number of safety exits for each fire zone shall be calculated and determined, and shall not be less than 2. When there are two or more adjacent fire compartments on the plane, each fire compartment may use a fire door on the firewall leading to the adjacent compartment as a second safety outlet, but there must be a safety outlet that leads directly to the outside. The threshold, steps and protrusions shall not be set up in the evacuation aisle, and shall be directly connected to the safety exit, and shall not pass through any room. The safety exit door shall be opened in the direction of evacuation, and no step shall be set up within 1.4 m near the door, so as to avoid tripping and trampling accidents during evacuation.

6. Strengthening fire safety management

First of all, unified management must be implemented. Each business owner inside the mall must obey the unified management of the mall fire protection department, sign "Certificate of Responsibility for Safety and Fire Prevention" step by step, and strictly implement the duty system of safety and fire prevention. Secondly, establish and improve the fire safety management system of firefighting. It shall be implemented and supervised by the person responsible for safety and fire prevention. Thirdly, the fire safety inspection shall be carried out according to "Provisions on the Administration of Fire Safety of Government Offices, Organizations, Enterprises and Institutions", and the inspection, repair and maintenance of fire-fighting facilities shall be done well according to its requirements, so as to ensure the integrity and good use of fire-fighting facilities. Finally, strengthen the fire safety training for managers and operators, so that everyone can understand and master the emergency measures, and will be able to alarm, will put out the first fire, will organize the evacuation of personnel.

9.1.3 Firefighting and Rescue in Underground Shopping Malls

The complex fire characteristics of underground commercial buildings determine that the firefighting and rescue tasks must be based on scientific organization and command and rich firefighting experience, and effectively adopt fire-fighting technology and tactics to improve the efficiency of fire accident disposal.

1. In the initial stage of fire, based on self-rescue, effectively control the fire

In order to avoid the serious fire of underground commercial buildings, it is very important to grasp the opportunity of extinguishing fire in the early stage. In the early stage of fire, the fume concentration in general buildings is low and the visibility is relatively high, so the principle of "self-rescue" should be adhered to.

(1) Organize personnel evacuation quickly to reduce casualties. Once a fire alarm system is passed or a fire is found, the management unit shall quickly assemble relevant personnel and organize mass evacuation and fire-fighting actions according to the evacuation plan or the actual situation of the fire site. Evacuation should choose the nearest and most convenient route, and evacuate in the direction of short distance and less danger. Evacuation should choose as many exits as possible, do a good job of personnel diversion, to avoid crowding stampede accidents.

(2) Make use of all kinds of fire-fighting facilities to effectively control the fire. In case of fire, the air-conditioning system shall be shut down immediately to stop the supply and prevent the

fire from spreading. Start the fume exhausting equipment, remove the fume from the fire, and improve the visibility of the fire. Open all kinds of fire prevention and extinguishing facilities, such as automatic fire extinguishing system, water curtain system and fire compartmentation system, to prevent the spread of fire. Use indoor fire hydrant system and fire-fighting equipment to organize relevant personnel to carry out fire-fighting operations.

2. Fire brigade adopts correct firefighting technology to carry out firefighting and rescue efficiently

When the fire brigade arrives at the fire scene, a fire fighting headquarters shall be set up immediately.

(1) Carry out fire investigation in a timely manner, and quickly deploy fire-fighting forces. Investigation methods can be adopted:

①External investigation. The fire site commander carefully understands the structure, scale, internal use of the building and the fire safety evacuation passages and internal fire protection facilities through external observation and inspection of the architectural drawings.

②Internal investigation. Under the premise of ensuring security, the reconnaissance team should go deep into the interior of the building and designate persons familiar with the building as guides to quickly find out the specific conditions inside the building.

③Instrumental test. Through the fire detector, gas detector, thermometer, detection robot and so on, the toxic gas composition and concentration, oxygen content in air, air temperature and so on are detected at the entrance of the building and deep inside.

(2) Fume exhausting should be carried out in time so as to make adequate preparations for firefighting and attack. The main fume exhausting methods are as follows:

①Natural fume exhausting by windows, evacuation doors, vertical shafts and air outlets.

②Use fixed or mobile fume control and exhaust equipment (fume exhaust fan or fume exhaust vehicle, etc.) to carry out mechanical fume exhaust.

③The water spraying or water spraying system in the building is used to exhaust fume, reduce temperature and reduce the concentration of fume in the fire site.

④According to the needs of the fire site, high-power foam can be used for fume exhaust.

(3) Evacuate trapped personnel and reduce casualties. In the early stage of the fire, more people are evacuating the building, rescue personnel should quickly organize forces, enter the internal guide personnel evacuation; When the fire is in the stage of intense combustion, rescue personnel shall be sent to the trapped personnels who cannot evacuate safely by themselves in a certain area to storm and rescue people under the cover of sufficient water flow.

3. Accurately grasp the operational opportunity and flexibly adopt fire-fighting tactics

In the firefighting of underground shopping malls, the fighters should be accurately grasped according to the specific position and fire situation, and the methods of internal attack, fire extinguishing, pouring fire extinguishing or sealing asphyxiation should be adopted flexibly.

(1) Attack and extinguish fire inside. In different stages of fire, according to the specific

situation of fire investigation and fume exhaust, multi-point internal attack should be organized and carried out at the right time, and the correct attack route should be chosen. In order to reach the fire area in the safest and fastest way, the fire should be destroyed step by step, as far as possible, in the interior of the building and close to the fire source.

(2) Pouring fire extinguishing. For underground caverns which are not suitable for internal attack, or some local lane ways and rooms of underground buildings, the method of pouring fire extinguishing agent into the underground can be used to extinguish the fire when no one is trapped.

(3) Sealing asphyxia. When the fire developed rapidly, The temperature is extremely high, When the fire fighters were seriously hampered in their intrusive operations, In order to shorten the battle time, the tactical method of sealing the entrance, air inlet and fume outlet can be adopted to close the fire area tightly and cut off the air source, so that the internal combustion area can be extinguished due to the lack of oxygen, i.e., the sealing asphyxiation method can be applied to extinguish the fire.

9.2　Disaster Prevention and Rescue of Underground Parking Lot

In the process of urban construction, the number of private cars is increasing, and the demand for underground parking is also increasing. Most underground parking lots not only have the basic functions of parking lots, but also shoulder the important task of civil air defense. Therefore, the disaster prevention and rescue of underground parking lot is very important.

9.2.1　Hazard of Underground Parking Lot

(1) Most of the causes of fire in underground parking lot are caused by the problems of vehicles themselves. In the absence of timely detection, the fire will spread and eventually lead to more serious consequences.

(2) Fire in the underground parking lot and the reason why there are combustibles in the car. Some vehicles are loaded with dangerous goods and parked in underground garages. If a fire breaks out in the parking lot when there are people in the parking lot, it is easy to cause casualties. Failure to evacuate safely could have even worse consequences.

(3) Underground parking is composed of many subsystems, including ventilation system, water system, power system and so on. In the event of a fire, these systems may be damaged, resulting in the temporary loss of part of the functions of the underground parking lot and, in serious cases, may endanger the ground.

(4) Underground parking spaces are relatively closed. Once there is a fire, people in the parking lot will panic, and then there will be a rush to drive away, which greatly increases the occurrence of car crashes. Not only that, in the event of fire will produce a large number of fume,

which will blur the driver's line of sight, leading to drivers in a hurry prone to accidents.

9.2.2 Fire Prevention and Control of Underground Parking Lot

1. Fire compartmentation

In order to avoid the spread of fire danger from one area to another, it is necessary to isolate the underground parking lot, control the scope of the fire, and avoid its expansion to cause greater impact when there is a fire in the underground parking lot. Specific measures are as follows:

(1) In the opening position of the connection between the garage and the passageway, two fire-proof coil brakes must be set up and controlled by the passageway and the garage respectively. In case of fire in the passageway or the garage, the coil brakes on the corresponding side shall be lowered to realize the function of fire isolation.

(2) According to the actual characteristics of the underground parking lot, the water curtain system should be used to divide it, and the intake air should be divided into several fire prevention zones. In each fire zone, a suitable number of fume staircases shall be provided, at least two or more, and at least one shall be guaranteed to have an independent evacuation function. In the stairwell length design, should also ensure that its walking distance within 60 meters. So that evacuation can be completed in the shortest possible time in the event of a fire.

2. Fume prevention and exhaust

Code for Fire Protection Design of Building (*GB 50016—2014*) stipulates that the building area of each fume prevention sub-area shall not be greater than 2000 m^2, and that the fume prevention sub-areas shall be divided by partition walls, fume prevention vertical walls or beams protruding not less than 0.5 m from under the roof. According to the natural conditions and the size of the underground parking lot, it is determined that the natural ventilation or mechanical ventilation should be chosen as the exhaust method. Longitudinal exhaust system can be adopted for underground parking lots with small change of channel slope and simple curvature. For the underground parking lot with large slope change and complicated curvature, the ventilation and exhaust system can only be constructed transversely inside the tunnel.

3. Fixed fire extinguishing and automatic alarm system

Code for Fire Protection Design of Garage, Motor Repair Shop and Parking Area (*GB 50067—2014*) stipulates that, except for open garages, several types of garages shall be equipped with automatic sprinkler systems and automatic alarm systems. The fire extinguishing system in the underground parking lot should be equipped with facilities linked with the fire alarm system, so as to mobilize the delivery equipment of the underground parking lot to extinguish the fire at the first time when the fire occurs, thus avoiding the lag of human activities. In some types of smaller parking lots cannot be equipped with automatic fire alarm system, in this case, the fire occurred without the help of the fire alarm, or fire alarm help function is relatively small, at this time, more attention should be paid to rely on their own strength to deal with the fire situation.

4. Safe evacuation

The evacuation exit of underground parking lot is divided into personnel safety exit and automobile evacuation exit, which should be set separately.

1) Setup of personnel safety exit

Code for Fire Protection Design of Building (GB 50016—2014) stipulates that the personnel safety exits of adjacent fire protection zones may be used as the second safety exits of this zone by the personnel safety exits of not less than one fire protection zone in the basement with direct access to the outdoor personnel safety exits. Therefore, each fire prevention zone of the underground parking lot is provided with a Class A fire door leading to the adjacent fire prevention zone as a "second safety outlet" in addition to an evacuation staircase.

2) Vehicle evacuation exit

The *Code for Fire Protection Design of Garage, Motor Repair Shop and Parking Area* (GB 50067—2014) stipulates that the car evacuation openings in the parking lot shall be arranged in different fire prevention sub-zones, and the total number of car evacuation openings in the whole parking lot shall be not less than 2. When setting up automobile evacuation openings, attention should be paid to the reasonable distance between the evacuation openings and the farthest parking space, and the fire prevention distance between automobile ramps and adjacent buildings should also be reasonably controlled (fire prevention distance $\geqslant 10$ m).

5. Strengthen the management level of fire safety

Large underground parking lot fire automatic fire control facilities are complete, fire safety management requirements are higher. The property management unit shall set up a professional maintenance management team, carry out regular inspection, eliminate unsafe factors in time, and ensure the integrity and good use of automatic fire-fighting facilities.

6. Strengthening personnel and vehicle management

Carry out rotation training for relevant parking owners so that they can master the necessary knowledge of extinguishing fire. At the same time, all irrelevant personnel shall not be allowed to enter, so as to prevent human sabotage. Carefully inspect the vehicles entering and leaving the vehicle to prevent "running, risking, dripping and leaking" of oil products.

9.2.3 Firefighting and Rescue in Underground Parking Lot

According to the characteristics of the underground garage fire, the fire commander should organize the fire reconnaissance carefully according to the specific situation of the fire, evacuate the personnel and vehicles under fire threat quickly, control the fire in time, and put out the fire quickly.

1. Keep abreast of the fire scene

Through external observation, interviewing insiders and organizing reconnaissance teams to conduct reconnaissance inside the underground garage, the fire scene commander promptly

Chapter 9 Disaster Prevention and Rescue of Other Underground Space

identified the following:

(1) Whether any person is trapped, the number of trapped persons, their location and the means of evacuation.

(2) Structure, scale, parking form, type and quantity of underground garage.

(3) Underground garage entrance and exit position, which can be used for vehicles to evacuate and extinguish the fire.

(4) Whether the fire place is the fire of the vehicle or the fire of the attached room; The size of the fire and the direction of its spread.

(5) Whether there are any fixed fire extinguishing facilities in the warehouse, whether they have been started, and how effective the fire control is.

2. Active evacuation of persons and vehicles

Firefighters should strengthen the evacuation of personnel and vehicles in the underground garage where the vehicle is on fire.

1) Evacuation

The ramp type underground garage and the compound type underground garage, because the underground garage has the personnel to stay. Therefore, after the reconnaissance confirms that there are people, it is necessary to organize forces to rescue and evacuate them quickly, and organize forces to control the fire so as to cover the rescue operation. Mechanical underground three-dimensional garage due to parking by mechanical equipment operation, underground garage without drivers, but pay attention to whether there is a garage staff trapped, confirm that someone should also do their best to rescue.

2) Vehicle evacuation

Evacuation and protection of vehicles is an important measure to prevent the spread of fire and reduce the loss of fire. Therefore, on the basis of the firefighting plan, the fire field commander shall decide the evacuation plan together with the person in charge of the unit, clarify the division of labor, and take separate responsibilities.

3. Rational organization of fume exhausting in the fire site

Fire in underground garage and high temperature of thick fume bring great disturbance and threat to fire extinguishing. Effective fume exhaust measures must be taken to enhance fume exhaust effectiveness.

(1) If there are daylighting windows (holes) on the ground floor of the garage, the daylighting windows (holes) can be opened or broken down for natural fume exhaust.

(2) The mechanical fume exhaust system of underground garage is used to start the fume exhaust fan and carry out the mechanical fume exhaust.

(3) Use mobile fume exhausting equipment, such as fume exhausting vehicle and fume exhausting machine to exhaust fume. However, when there are more than two entrances and exits in the garage, the following air outlets shall be used as fume exhaust outlets and the other air

outlets as supply outlets.

(4) Depending on the needs of the fire site, the fume can be exhausted by spraying water or high-power foam.

4. Flexible use of firefighting methods

In order to extinguish the underground garage fire, the fire should be extinguished by means of internal attack, sealing or pouring according to the specific position and situation of the fire.

1) Attack and extinguish fire inside

Determine the attack route, quickly lay the water belt from the outside, from the upwind direction of the entrance and exit to the burning area, foam or spray water to extinguish the fire. When extinguishing the fire, the vehicles threatened by the fire shall be protected by water injection as far as possible, with the emphasis on cooling the fuel tank, preventing the fuel tank from exploding and expanding the spread of the fire.

2) seal or pouring fire extinguish

If the fire develops rapidly, the ignition area can be sealed tightly, the oxygen supply can be cut off, and the internal combustion can be extinguished automatically because of lack of oxygen under the condition of no direct internal attack and no effective external attack. For small underground garage fire, if the fire is fierce, high temperature, can not be in-depth underground attack extinguishing, can be poured into the underground garage high-power foam extinguishing.

9.3 Mine Disaster Prevention and Rescue

China's coal mine are widely dispersed, geological conditions vary widely, The number of coal mines is large and the conditions and levels are uneven. Coal mine fires and accidents caused by them occur from time to time. Mine machine fires seriously threaten the safety of mine production and threaten the lives and health of workers. At the same time, because 70% of the energy in China's industrial production depends on coal, the occurrence of mine fires has made tens of thousands of tons of coal sealed, frozen and burned.

9.3.1 Classification and Hazard of Mine Fire

1. Classification of mine fires

Mine fire must have three basic conditions: heat, oxygen, combustible burning.

According to the cause of spontaneous combustion, the mine fire can be divided into internal and external causes. The internal cause of fire is the accumulation of fire caused by the change of physical and chemical properties of coal and storage heat under certain conditions and environment. External fire refers to the fire caused by blasting, gas, open fire, improper operation of mechanical and electrical equipment and other reasons.

Chapter 9 Disaster Prevention and Rescue of Other Underground Space

According to the combustion state, the mine fire can be divided into smoldering fire and open fire. When the air permeability is poor and the combustion place is very anoxic, smoldering will occur, and smoldering will often produce a large number of toxic gases. When the oxygen content is high, there will be a long-term flame combustion, full combustion, for open fire combustion.

2. Hazard of mine fire

Mine fire causes casualties and property losses, including direct casualties caused by high temperature of fire source, toxic casualties caused by high temperature toxic and harmful gases in the leeward direction, gas explosion induced by mine fire, resulting in greater disasters, high temperature caused by mine fire destroys roadway support, burns coal, and forms safety hidden dangers. Mine fires also affect the normal production order, making the environment deteriorate, production decline, or even stop production.

9.3.2 Mine Fire Prevention and Control

1. Prediction of mine fire

(1) Identify the spontaneous combustion tendency of coal. Chromatographic oxygen uptake identification method can be used, which can be used to detect the ability of coal to absorb oxygen at low temperature (oxygen content, velocity).

(2) Predict the fire in each layer. There are two main aspects: one is to master the use of forecasting instruments and devices. Mainly used in our research and development of more sensitive and reliable detection indicators and new indicators to adapt to the instrument or device, as well as sensitive components. Second, master the use of forecast indicators. As for detection index, coal spontaneous combustion can be divided into three stage by using CO, C_2H_4 and C_2H_2; slow oxidation stage when only 10^{-6} grade CO occurs in mine air flow; Accelerated oxidation occurs when CO and C_2H_4 of grade 10^{-6} are present. When CO, C_2H_4 and C_2H_2 of grade 10^{-6} appeared, it was the intense oxidation stage, at this time, open fire would appear. The application of three indexes can not only predict the fire, but also distinguish its stages, so different fire prevention measures can be taken.

(3) Do well the mechanical and electrical equipment and chamber fire detection system. In order to predict and prevent the fire accident of belt conveyor or electromechanical chamber, we should master the application of automatic fire extinguishing system and fire monitoring system.

2. General measures for mine fire prevention

(1) Adopt non-combustible material to support. When drilling along coal seam, shaft, adit and bottom-hole depot, arch must be laid; Non-combustible materials must be used for support when drilling in rock strata.

(2) Establish a fire-fighting material database. Each mine must store fire-fighting materials and tools, and establish a number of fire-fighting warehouses, fire-fighting warehouse materials to

be regularly inspected and replaced.

(3) Set fire doors.

Both the air inlet and the air inlet adit shall be equipped with fire-proof iron doors, which shall be tightly covered and easy to close. Two easy-to-close iron doors or iron-clad fire doors on wooden boards shall be installed at the connection between the intake shaft and the various transverse underground depots.

(4) Set up firefighting pool and underground firefighting duct system.

Each mine must be equipped with a fire-fighting pool and underground fire-fighting piping system on the ground. A water pump shall be installed near the firefighting pool, and its head and discharge shall be specified in the design of mine firefighting equipment. In addition to the surface fire water tank, the upper transverse or production level water tank can also be used as the fire water tank in the deep transverse mine.

3. External fire prevention

Prevention of external fires should start with the elimination of open fires and electric sparks. The main measures are as follows:

(1) Safety explosives shall be used in gas mines and safety regulations shall be observed when firing guns.

(2) Correctly select, install and maintain the electrical equipment to ensure that the wiring is in good condition and prevent short circuit and overload from generating sparks.

(3) It is strictly forbidden to use bulbs for heating and electric furnaces in underground. Downhole and wellhead room shall not be engaged in electric welding, gas welding or blowtorch welding.

4. Prevention of internal fire

(1) The requirements of correct selection of development and mining methods to prevent spontaneous combustion fire for development and mining are: minimum coal seam exposed surface, maximum coal mining capacity, fastest step-back speed and easy isolation of mining area.

(2) Adopt correct ventilation measures. The first is to select reasonable ventilation system in mining area. Combining the development plan and the mining sequence, the reasonable ventilation of the mining area is selected. Second, implement ventilation in the wind zone. Divided ventilation is reasonable, which can reduce the total resistance of the mine, expand the ventilation capacity of the mine, and easy to adjust the air volume. At the same time, it is also convenient to stabilize air flow and isolate the fire area during the fire.

(3) Do well in preventive grouting. Preventive grouting is to send grouting materials to areas prone to spontaneous combustion by means of grouting equipment, which plays a role of fire prevention. The grouting material can be slurry or tailings. Inhibitor can be used for grouting fire prevention in the mine area lacking soil and water.

Chapter 9 Disaster Prevention and Rescue of Other Underground Space

9.3.3 Mine Firefighting and Rescue

1. Basic Measures for Fire Extinguishing and Rescue

(1) Cut off the power supply in the fire area after the fire, so as to prevent the ambulance men from electric shock and methane explosion in the course of dealing with the fire.

(2) Those who are evacuated from the disaster area immediately after the fire and who are threatened in case of gas explosion.

(3) Actively rescue the people in distress and take measures to prevent the fume from spreading to the crowded areas.

(4) A special person shall be set up to inspect the change of gas and air flow, so as to prevent the air flow from harming people in reverse.

(5) The ambulance team shall find out the place, scope and cause of the fire as soon as possible.

2. Firefighting and rescue measures at different locations

1) Near the air inlet

In case of fire near the air inlet, the following measures should be taken:

(1) The ventilation fan should carry out backwind;

(2) Close that underground, and closing the fire-proof iron door of the air inlet;

(3) Safely evacuate that underground miners accord to the disaster avoidance route;

(4) Adopt suitable methods to extinguish the fire and prevent the fume flow from entering the well.

2) Shaft bottom

The following measures should be taken in case of fire in the shaft bottom:

(1) The ventilation fan of the whole mine should carry out the reverse wind;

(2) Safely evacuate that underground miners accord to the disaster avoidance route;

(3) If the water source of the shaft bottom is sufficient, extinguish the fire directly with water;

(4) In order to reduce the supply near the fire in the shaft bottom, the temporary sealing method is adopted to prevent the fire from spreading;

(5) If it is a mine with central parallel ventilation, short circuit of intake and return air can be adopted to discharge the fume flow.

3) Air intake shaft

In case of fire in intake shaft, the following measures should be taken:

(1) The ventilation fan in the whole mine should carry out reverse wind;

(2) Safely evacuate the underground miners according to the disaster avoidance route;

(3) When the anti-wind effect is not good, close the fire-proof iron door to reduce the oxygen supply;

(4) If the fire is not serious in the inclined shaft, enter the shaft to put out the fire. If the fire is large, wait until the counter-wind is effective before entering the wellbore to put out the fire;

(5) When the shaft is on fire, the high foam extinguisher can be used on the ground to extinguish the fire.

4) Return air shaft

In case of fire in return shaft, the following measures should be taken:

(1) Do not change the direction of air flow;

(2) Control that air inlet fire door, stopping part of the fan, and reducing the air volume;

(3) Under the condition of multi-air shaft and multi-fan ventilation, the main fan of the return air shaft in the fire area cannot stop the ventilation;

(4) When the fire is enlarged, evacuate the underground miners safely according to the disaster avoidance route.

5) Main chamber

The following measures should be taken in case of fire in the main chamber:

(1) When the electromechanical chamber is on fire, the power supply should be cut off immediately, and then the field tools should be used to extinguish the fire;

(2) Explosive materials shall be transported out immediately when the powder depot is on fire. If the temperature is too high to be transported out, close the fire doors and evacuate to a safe place;

(3) When the winch house is on fire, fix the mine truck below the fire source to prevent the rope of the mine truck from being burnt off and running away to injure people;

(4) When the battery garage is on fire, the power supply shall be cut off first, and then the battery shall be transported out. If the fire is very big and difficult to put out, close the fire-proof iron doors and then adopt positive measures to put out the fire.

6) Ventilation gallery

In case of fire in ventilation gallery, the following measures should be taken:

(1) In case of fire in inclined intake gallery, short circuit of fume flow and counter-wind should be taken;

(2) In case of fire in transverse passage, the supply should be increased or decreased according to the change of gas quantity.

7) Goaf

The fires in goaf are usually caused by internal factors. The following measures should be taken:

(1) It is difficult to extinguish the fires in goaf by direct extinguishing method, usually by isolation method;

(2) If there are more air leaks in goaf and the effect of isolating fire extinguishing method is not good, the method of pouring mud water can be used to extinguish fire;

(3) That method of inject inert gas can also be used for extinguish the fire;

(4) Equalization of pressure can also be used to extinguish the fire.

8) Coal Mining Face

The following measures should be taken in case of fire in coal mining face:

(1) When the fire source can be approached, water and fire extinguisher should be used to extinguish the fire directly; extinguish the fire at a distance using a high foam machine when the fire source cannot be approached;

(2) When gas burns in coal mining face, use dry powder fire extinguisher or sand, rock powder, soil, etc. to extinguish fire;

(3) When the fire extinguishing effect cannot be achieved from the air inlet side, the fire extinguishing effect can be achieved from the air return side;

(4) If the fire cannot be extinguished directly or there is danger of gas explosion, the fire area can be sealed off, and the fire can be extinguished by asphyxiation.

3. Extinguishing methods

1) Positive methods for extinguishing fire

In general, active methods should be used as far as possible to extinguish fires. The specific methods include: extinguishing fire with water; Extinguish fire with inert gas; Extinguish fire with sand, rock powder, soil, etc. Extinguish fire with a fire extinguisher; Extinguish fires by digging up burners.

2) Isolation method for extinguish fire

Isolation fire extinguishing method is a kind of fire extinguishing method when there is no fire extinguishing equipment and personnel, or direct fire extinguishing method can not achieve the desired effect, or personnel are difficult to access the fire area. A method for quench that fire area includes close the fire area, cutting off the air leading to the fire area, applying pressure equalization technique, and pouring mud water, such as inert gas, into the fire area to quench the fire area.

3) Comprehensive method for extinguish fire

The comprehensive method of extinguishing fire is the combination of active method and isolated method of extinguishing fire. First, the isolation method is used to speed up the fire extinguishing speed, and then the closed wall is opened to extinguish the fire in a positive way.

9.4 Disaster Prevention and Rescue of Civil Air Defense Engineering

Civil air defense engineering is very important for a country, once a fire, because of its secrecy, not suitable for evacuation and other reasons will cause significant economic losses, a large number of casualties. Therefore, the disaster prevention and rescue of civil air defense engineering is very important.

9.4.1 Characteristics of Fire in Civil Air Defense Works

1. Hazard of fire fume

When a fire breaks out in a civil air defense project, as that underground ventilation effect is poor, the toxic and high-temperature fume accumulated in a short time cannot be effectively discharged. High concentration of toxic fume will not only help the fire to expand its burning range and speed up its spreading speed, but also bring great threat to the safety of people evacuated from underground civil air defense buildings. It may even bring great rescue difficulties to the firefighters who come to rescue because of low visibility, high temperature, fume poison and other reasons.

2. Fire susceptibility

Usually, underground civil air defense works in peacetime are mostly used as crowded places, such as: public entertainment, cinema, and so on. Its internal furniture, decoration and other combustibles are relatively more. In addition, civil air defense works are underground. Compared with the moist ground, the insulation layer of the electrical equipment and lines in this environment is easy to corrode, and the electrical lines in the absence of insulation layer protection occur local short circuit fire or local contact resistance is too large, then the insulation layer of the electrical lines become combustible, which forms a hidden danger of electrical fire.

3. Fast spread of fire

When a fire breaks out, most of the fume and heat of the ground floor building escapes through the doors and windows of the building. Compared with the above-ground buildings, because of its small aperture to the ground and small internal space, plus the air defense works are surrounded by dirt. In the event of a fire, the heat released by combustion will not be easily dissipated, which makes the indoor temperature of this kind of building rise rapidly in a short time, and the high-temperature fume which cannot be dissipated will further heat other combustibles through radiation or convection transfer, thus further accelerating the spread of the fire, resulting in an out-of-control situation.

4. Concealment of fire

Because of the closeness and limitation of its own buildings, it is difficult for civil air defense engineering to find fire in time and organize escape, rescue and other actions. There is plenty of ancillary space for civil air defense works, strong concealment, poor connecting signals, and the closure of their structures is difficult to achieve effective connecting between each other. Once a fire occurs, it is difficult to find out in time. Even if a fire is found in time, it is also difficult to quickly use conventional means of connecting to report the fire, organize personnel evacuation, and direct the fire fighters into the underground civil air defense buildings to carry out firefighting and rescue.

5. Difficulty in evacuation

First of all, the passageway of civil air defense engineering is narrow, and the exit is relatively small, especially because it is buried deep in the ground. Once there is a fire, the evacuation of personnel is more difficult. Secondly, the visibility of underground civil air defense works is low. Usually, artificial lighting is used in underground civil air defense works, and artificial lighting is significantly weaker than natural lighting in illumination. Finally, a large number of fume produced by the fire affected the evacuation speed.

9.4.2 Fire Prevention and Control of Civil air Defense Works

1. Fire and fume compartmentation

After a fire broke out in the civil air defense project, the spread of the fire will be difficult to control in the first place. So, at design time, it is necessary to strictly comply with the requirements of the State *Code for Fire Protection Design of Civil Air Defense Buildings*, Fire prevention and fume prevention design and separation shall be carried out for fire prevention and fume prevention zones, so as to facilitate fire prevention facilities such as firewalls, fire doors, fire curtains, fire windows, fire valves and fume prevention vertical walls to control the spread of fire, discharge toxic fumes, grasp the evacuation time, ensure the safety of personnel and reduce property losses. In addition, when fire-proof rolling shutter doors are used for fire-proof separation in the evacuation passageway, fire-fighting facilities such as fume-proof vertical walls and fume-exhausting fans should be added to control fire and prevent fume.

2. Installation of fixed fire fighting facilities

First of all, an intelligent automatic alarm system is set up to improve the speed, accuracy and reliability of the fire alarm system. Secondly, according to the special situation of fire in civil air defense project, an effective automatic fire extinguishing system is set up, such as diaphragm open large water droplet rain shower fire extinguishing system.

3. Setup of light-emitting evacuation indication signs

In the event of a fire in the civil air defense project, because of its poor ventilation, plus the blackout, thus greatly affecting visibility, It can be considered to add some auxiliary emergency evacuation indication signs on the ground and the lower part of the evacuation passage, which rely on fluorescent materials to absorb and emit light. These materials are durable and inexpensive. Because of their short absorption time and long discharge time, they can provide effective evacuation indication within 20-30 minutes after the power cut-off of the fire to ensure timely, effective and orderly evacuation of personnel.

4. Division of responsibilities, training and drills

Achieving hierarchical and decentralized management of staff, clarify the responsibilities and tasks of fire control for each post and job type, we should also organize new personnel to conduct fire-fighting training and conduct fire-fighting drills in a timely manner. All personnel in the site can be familiar with the environment around their own location, understand that once a fire "where to run, how to run, how to put out the initial fire, how to alarm", with the requirements of "four abilities of fire safety", so that personnel can evacuate in time, escape from self-rescue and put out the initial fire.

5. Be familiar with shelter or walkways

Temporary fire safety areas such as shelter places or walkways are often set up in civil air defense works. In case of fire, if the personnel in the site cannot escape in time, during the fire drill, the users in the site shall be informed, consider temporary safe areas such as shelters or corridors where people who cannot escape in time can choose to wait for firefighters to come to the rescue or escape to other relatively safe neighboring areas through the corridors to provide valuable time for themselves and rescuers.

9.4.3 Firefighting and Rescue of Civil Air Defense Works

In peacetime, civil air defense works are often used as underground shopping malls, underground car parks and public entertainment places, and their firefighting and rescue measures are almost the same as those of underground shopping malls, underground car parks and public entertainment places. The firefighting and rescue of civil air defense engineering can be divided into personnel, vehicle evacuation and fume suppression.

1. Evacuation of personnel and vehicles

1) Evacuation

Once the fire alarm system in the civil air defense project discovers fire, the staff in the civil air defense project quickly organizes the masses to evacuate along the various safety exits. Then wait until the firefighters arrived at the scene, again organize the strength to guide the majority of people evacuated from the civil air defense project. If a small number of people cannot be evacuated from the civil air defense project because of the serious fire, rescue personnel shall rescue them under the cover of certain fire control measures.

2) Vehicle evacuation

Firefighters evacuate and protect vehicles to prevent the spread of the fire, but also to make effective space for emergency rescue, while guiding the fire and ambulance vehicles into the reasonable layout of the site, effective operation.

2. Exhaust fume

1) Exhaust fume

If the lighting window is set in the civil air defense project, the lighting window can be used for natural fume exhaust; Start the fume exhauster or the fume exhauster to exhaust fume by using the mechanical fume exhausting equipment in the civil air defense project; Exhaust fume with sprayed water or high-power foam.

Chapter 9 Disaster Prevention and Rescue of Other Underground Space

2) Extinguishing fire

Once a fire is discovered, the automatic sprinkler system in the underground projects starts to put out the fire. When a fire is first discovered, fire extinguishers can be used by the staff of civil air defense engineering to extinguish the fire. If the development of the fire is small, it can be extinguished by internal attack, laying water belts from the outside, entering the interior to foam or spray water to extinguish the fire. For some vehicles affected by the fire, water jetting protection shall be carried out; If the fire develops rapidly and cannot be extinguished by internal attack, it can be poured or sealed from the outside.

 **Exercise**

9.1 Briefly describe the risk of fire in underground shopping malls.

9.2 What are the requirements for fire prevention zones of underground shopping malls?

9.3 What firefighting methods are available in underground shopping malls? Under what conditions?

9.4 Briefly describe the fire hazard of underground parking lot.

9.5 What are the provisions for the installation of car evacuation gates in underground parking lots?

9.6 What is the internal cause of mine fire? What is an external fire?

9.7 Briefly describe the prevention means of mine internal fire and external fire.

9.8 Brief description of firefighting and rescue measures for civil air defense engineering.

 Key vocabulary:

accelerated oxidation　加速氧化
coal mining　煤矿开采
coal seam　煤层
fire door　防火门
goaf　采空区
non-combustible and uninflammable interior decoration　非燃难燃化内部装修

References

[1] Yang Lixin. Ventilation technology for modern tunnel construction [M]. Beijing: People's Communication Publishing House, 2012.

[2] 中华人民共和国行业标准. TB 10003—2016 铁路隧道施工规范[S]. 北京:中国铁道出版社,2016.

[3] 中华人民共和国交通行业标准. JTG F60—2009 公路隧道施工技术规范[S]. 北京:人民交通出版社,2009.

[4] 中华人民共和国标准. GB 16423—2006 金属非金属矿山安全规程[S]. 北京:中国标准出版社,2006.

[5] 中华人民共和国标准. GBZ 1—2010 工业企业设计卫生标准[S]. 北京:人民卫生出版社,2010.

[6] 中华人民共和国标准. GBZ 2.1—2019 工作场所有害因素职业接触限值 第一部分:化学有害因素[S]. 北京:人民卫生出版社,2019.

[7] 中华人民共和国标准. GBZ 2.2—2019 工作场所有害因素职业接触限值 第二部分:物理因素[S]. 北京:人民卫生出版社,2019.

[8] 中华人民共和国煤炭行业标准. MT/T634—2019 煤矿矿井风量计算方法[S]. 北京:应急管理出版社,2019.

[9] 中华人民共和国标准. GB/T 9900—2008 橡胶或塑料涂覆织物导风筒[S]. 北京:商务出版社,2008.

[10] 中华人民共和国标准. GB/T 15335—2019 风筒漏风率和风阻的测定方法[S]. 北京:中国标准出版社,2019.

[11] 中华人民共和国化工行业标准. HG/T 2580—2008 橡胶或塑胶涂覆织物拉伸长度和拉断伸长率的测定[S]. 北京:化工出版社,2008.

[12] 中华人民共和国交通行业标准. JTG/T D70/2-02—2014 公路隧道通风设计细则[S]. 北京:人民交通出版社,2014.

[13] Gao Bo, Wang Yingxue, Zhou Jiamei. Subway [M]. Chengdu: Southwest Jiaotong University Press, 2011.

[14] Yu Huaqian, Chen Chunguang, Mai Jiting. Engineering fluid mechanics [M]. Chengdu: Southwest Jiaotong University Press, 2013.

[15] Fu Gang, Wang Cheng. Tunnel ventilation and lighting [M]. Wuhan: Wuhan University Press, 2015.

[16] Liu Jian. Safety and lighting of tunnel ventilation [M]. Chongqing: Chongqing University Press, 2015.

[17] 中华人民共和国行业标准. TB 10068—2010 铁路隧道运营通风设计规范[S]. 北京:中国铁道出版社,2010.

[18] Chuan Fu. My opinion on the development direction of ventilation mode in railway tunnel operation [J]. Railway Architecture, 1997 (05): 32-33.

［19］中华人民共和国标准. GB 50157—2013 地铁设计规范［S］. 北京:中国建筑工业出版社, 2013.

［20］中华人民共和国标准. GB 50019—2003 采暖通风与空气调节设计规范［S］. 北京:中国标准出版社, 2003.

［21］中华人民共和国标准. GB 50225—2005 人民防空设计规范［S］. 北京:中国建筑标准设计研究院, 2005.

［22］中华人民共和国标准. GB 50038—2005 人民防空地下室设计规范［S］. 北京:中国建筑标准设计研究院, 2005.

［23］中华人民共和国行业标准. JGJ 48—2014 商店建筑设计规范［S］. 北京:中国建筑工业出版社, 2014.

［24］中华人民共和国标准. GB/T 18883—2002 室内空气质量标准［S］. 北京:中国标准出版社, 2002.

［25］中华人民共和国标准. GB 16423—2006 金属非金属矿山安全规程［S］. 北京:中国标准出版社, 2006.

［26］Wan Chuanqi, Long Yu. Ventilation and air-conditioning design of underground shopping malls［J］. Chinese Folk Dwellings (late issue), 2014 (06): 62.

［27］Liu Yakun. Ventilation and air conditioning in underground shopping malls［J］. Building Thermal Energy Ventilation and Air Conditioning, 2002 (05): 39-40.

［28］Wu Danjun. Ventilation and air conditioning design of underground shopping malls［J］. Energy Saving Technology, 2000 (05): 10-11.

［29］Jiang Dong. Experience of HVAC design in underground shopping malls［J］. HVAC, 1999 (04): 54.

［30］Wang Wenkui, Jiao Youfen. Air conditioning, ventilation and smoke control in large underground shopping malls［J］. Forestry Science and Technology Information, 2005 (02): 60-61.

［31］Zhang Miao. Investigation and analysis of air conditioning ventilation and smoke control in large underground shopping malls［J］. Chinese Folk Dwellings (late issue), 2012 (06): 80.

［32］Wang Ying. Ventilation design based on underground parking lot［J］. Heilongjiang Water Resources Science and Technology, 2010, 38 (01): 107-108.

［33］Liang Xiaojun. Design of ventilation and smoke control system for underground parking lot［J］. Science and Technology Pioneer Monthly, 2005 (08): 152-153.

［34］Miao Changzheng. Ventilation and smoke control in underground parking lots［J］. Heilongjiang Science and Technology Information, 2011 (15): 297.

［35］Liu Li. Discussion on ventilation and smoke exhaust design of underground garage［J］. China New Technology and Products, 2012 (03): 179.

［36］Ren Zengyu. Discussion on mine ventilation technology and optimization design of ventilation system［J］. Heilongjiang Science and Technology Information, 2010 (12): 47.

[37] Zou Wei. Discussion on optimization design of mine ventilation technology and ventilation system [J]. Energy and Energy Conservation, 2014 (08): 33-34+39.

[38] Liu Zhiqiang. Summary of ventilation design for civil air defense engineering [J]. Science and Technology Wind, 2011 (05): 143. 2017-09-02.

[39] Li Zhanchu. Some problems in ventilation design of civil air defense engineering [J]. Refrigeration, 2006 (01): 77-79.

[40] 中华人民共和国行业标准. TB 10020—2017 铁路隧道防灾疏散救援工程设计规范[S]. 北京:中国铁道,2017.

[41] Wang Zaibian, Guo Chun, Yang Qixin. Disaster Prevention and Rescue Technology of Expressway Tunnels and Tunnel Groups [M]. Beijing: People's Communications Publishing House, 2010.

[42] 中华人民共和国标准. GB 50490—2009 城市轨道交通技术规范[S]. 北京:中国建筑工业出版社,2009.

[43] 上海市工程建设规范. DG/T J08-109—2017 城市轨道交通设计规范[S]. 上海,同济大学出版社,2017.

[44] 上海城市轨道交通网络建设标准化技术文件. STB/ZH-000001—2012 上海城市轨道交通工程技术标准(实行)[S]. 同济大学出版社.

[45] 中华人民共和国标准. GB 50166—2019 火灾自动报警系统施工及验收规范[S]. 北京:中国计划出版社,2019.

[46] 中华人民共和国标准. GB/T 50314—2015 智能建筑设计标准[S]. 北京:中国计划出版社,2015.

[47] 中华人民共和国标准. GB 50440—2007 城市消防远程监控系统技术规范[S]. 北京:中国计划出版社,2007.

[48] 中华人民共和国标准. GB 50174—2008 电子信息系统机房设计规范[S]. 北京:人民出版社,2009.

[49] 中华人民共和国行业标准. JGJ 16—2008 民用建筑电气设计规范[S]. 北京:中国建筑工业出版社,2008.

[50] 中华人民共和国标准. GB 16806—2006 消防联动控制系统[S]. 北京:中国标准出版社,2006.

[51] Zhang Zhirui. Discussion on fire risk analysis and fire prevention measures of underground shopping malls [J]. Architectural Knowledge, 2016, 36 (02): 288.

[52] Chen Peng. Performance-based fire protection design and study of underground shopping malls [D]. Xi'an University of Architecture and Technology, 2014.

[53] Li Bin. Discussion on fire prevention and evacuation of underground shopping malls [J]. Entrepreneur World: Theoretical Edition, 2010 (07): 224-225.

[54] Bai Weiming. Problems and preventive measures of fire prevention in underground shopping malls [J]. Science and Technology Wind, 2010 (04): 145.

[55] Zhang Shibin. Discussion on fire prevention and extinguishing measures of underground

garage [J]. Law and Society, 2015 (33): 212-213.

[56] Liu Guanchen. Study on fire process and fire control measures of underground garage [J]. Heilongjiang Science and Technology Information, 2014 (31): 2.

[57] Wang Yongxi. Fire fighting in underground garage [J]. Fire Fighting Technology and Product Information, 2009 (08): 31-33.

[58] Huang Ligang. Prediction of mine fire and technical measures for fire prevention [J]. Science Technology and Enterprises, 2011 (15): 29-30.

[59] Kou Haiping. Mine fire prevention methods [J]. Science and Technology Information, 2010 (08): 302 +300.

[60] Zhang Xiaojun. Mine fire prevention measures [J]. Journal of Shanxi Coal Management Cadre College, 2013, 26 (03): 45-46.

[61] Ma Junling. Causes and control measures of mine fires [J]. Science and Technology Information, 2010 (33): 367.

[62] Yu Haoyu. Fire protection measures for civil air defense engineering [J]. Science and Technology Prospect, 2015, 25 (20): 265.

[63] Song Yang. Discussion on fire prevention measures for civil air defense engineering [J]. China Urban Economy, 2011 (14): 289-290.

[64] 中华人民共和国标准. GB 6499—2012 危险货物分类和品名编号[S]. 中国标准出版社, 2012.

[65] 中华人民共和国标准. GB 50098—2009 人民防空工程设计防火规范[S]. 北京: 中国计划出版社, 2009.

[66] 中华人民共和国标准. GB 50222—2015 建筑内部装修设计防火规范[S]. 北京: 中国计划出版社, 2015.

[67] 中华人民共和国标准. GB 50016—2014 建筑设计防火规范[S]. 北京: 中国计划出版社, 2014.

[68] 中华人民共和国标准. GB 50067—2014 汽车、修车库、停车场设计防火规范[S]. 北京: 中国计划出版社, 2014.